D1130350

Authority, Liberty, and Automatic Machinery in Early Modern Europe

Johns Hopkins Studies in the History of Technology
Thomas P. Hughes, general editor

Authority, Liberty & Automatic Machinery in Early Modern Europe

Otto Mayr

The Johns Hopkins University Press

Baltimore and London

Printed in the United States of America

The Johns Hopkins University Press
701 West 40th Street
Baltimore, Maryland 21211
The Johns Hopkins Press Ltd., London

The paper in this book is acid-free and meets the guidelines for permanence and durability of the Committee on Production Guidelines for Book Longevity of the Council on Library Resources.

Library of Congress Cataloging-in-Publication Data

Mayr, Otto.
Authority, liberty, and automatic machinery in early modern Europe.

(Johns Hopkins studies in the history of technology)
Bibliography: p.
Includes index.
1. Technology and civilization—History. 2. Europe—Intellectual life.
3. Great Britain—Intellectual life. 4. Authority. 5. Liberty.
I. Title. II. Series.
HM221.M34 1986 303.4'83'094 85-15460
ISBN 0-8018-2843-0 (alk. paper)

To Louise

Contents

List of Illustrations xi

Acknowledgments xiii

Introduction xv

I Authoritarian Systems

1 The Mechanical Clock, Its Makers and Users 3
2 The Rise of the Clock Metaphor 28
Clock Metaphors as Evidence 28
Late-Gothic Beginnings 30
Flowering in the High Baroque 41
3 The Clockwork Universe 54
The Mechanical Philosophy 54
Descartes 62
Continental Mechanists 67
British Mechanists 81
Determinism versus Free Will 92
4 The Clockwork State 102
5 The Authoritarian Conception of Order 115
6 Rejection of the Clock Metaphor in the Name of Liberty 122

II Liberal Systems

7 Imagery of Balance and Equilibrium 139
8 Attraction and Repulsion 148
9 Self-balancing Political Systems 155
10 Self-regulation in Economic Thought 164
11 Self-regulation and the Liberal Conception of Order 181
12 Self-regulating Mechanisms in Practical Technology 190

Notes 201

Illustration Credits 259

Index 261

Illustrations

1–1 Nuremberg fifteenth-century wall clock 6
1–2 (A and B) Rooster automaton of the first Strasbourg monumental clock, 1354 7
1–3 Planetary clock of Giovanni de' Dondi, Padua, between 1348 and 1364 8
1–4 Fusee and spring barrel 9
1–5 Table clock, presumably a masterpiece, Augsburg, 1600 11
1–6 (A and B) Table clock by Steffen Brenner, Copenhagen, 1558 12
1–7 Second astronomical clock of the Strasbourg Cathedral at its completion in 1574 13
1–8 (A and B) After the Horological Revolution: Two bracket clocks 16
1–9 Portrait of a young man, circa 1590 18
1–10 Portrait of Cardinal Richelieu by Philippe de Champaigne, 1622 19
1–11 Portrait of Don Justino de Neve by Bartolomé Esteban Murillo, circa 1650 20
1–12 Astronomical clock with automata displays, modeled after the Strasbourg monumental clock 22
1–13 Program-controlled mechanical pipe organ of the table carriage "Minerva" 23
1–14 Tower clock, Marburg, seventeenth century 23
1–15 Planetary clock by Eberhard Baldewein, Marburg, 1558–62 24
1–16 Three automaton table carriages 25
2–1 "Sapientia" amidst a selection of timepieces 33
2–2 "Temperantia" adjusting a mechanical clock 36
2–3 "Temperantia," engraving by Pieter Bruegel the Elder, circa 1560 37

3–1 Sphere of the planet Mercury, from a sixteenth-century commentary on Georg Peurbach's *Theoricae novae planetarum* (1454) 60
10–1 Isaac Gervaise's "natural ballance of trade" (1720) 167
10–2 Isaac Gervaise's theory of the equilibrium between the desire for wealth and the love of ease 168
10–3 David Hume's self-regulating mechanism for international monetary distribution 171
10–4 Adam Smith's general theory of supply and demand—1: price as a function of demand minus supply 176
10–5 Adam Smith's general theory—2: supply as a function of price minus cost 177
10–6 Adam Smith's general theory—3: the complete system 177
10–7 Adam Smith's theory of the supply and demand of labor 178
10–8 Adam Smith's theory of the self-balancing nature of the rewards of all occupations 180
12–1 Float-level regulator by Hero of Alexandria, circa A.D. 60 191
12–2 Hero's float-level regulator, as depicted in Federigo Commandino's 1575 Latin edition 191
12–3 Float-level regulator in the feed-water supply tank of the steam pump by Thomas Savery, circa 1700 192
12–4 Cornelis Drebbel's thermostatically regulated chicken incubator, circa 1620 193
12–5 Cornelis Drebbel's thermostatically controlled alchemical furnace, circa 1620 193
12–6 Pressure cooker with pressure regulator by Denis Papin, 1681 194
12–7 Speed regulation scheme for mills by G. W. Leibniz, 1686 195
12–8 Speed regulator for windmills by Thomas Mead, 1787 196
12–9 The "fan-tail" 197
12–10 Centrifugal governor on a Watt steam engine, circa 1790 198

Acknowledgments

I wrote most of this book while I was a curator at the then National Museum of History and Technology of the Smithsonian Institution, Washington, D.C. During this time my work benefited not only from the Smithsonian's general policy of encouragement and support of research but also specifically during 1970–71 from a Smithsonian research grant.

It is difficult to give adequate credit for all the help that came to me in one form or another. The crucial impulses and stimuli one receives from the general intellectual environment and through informal interaction with fellow scholars are impossible to acknowledge. Teachers who principally shaped my standards, outlook, and approach to the history of technology were Professor Friedrick Klemm (1914–83) and Professor Joachim Otto Fleckenstein (1914–80), both of the Technical University of Munich. Although I never entered his classroom, I would also like to claim as my teacher Professor Lynn White, Jr., of UCLA, simply because his *Medieval Technology and Social Change* was the prototypical book for my generation of researchers.

Many friends, colleagues, and institutions helped me in more specific ways. Concerning conceptual questions, I learned a great deal in conversations with Dr. Paul Forman, Smithsonian Institution, Professor Francis C. Haber, University of Maryland, Dr. Klaus Maurice, Deutsches Museum, Munich, Dr. Robert Multhauf, Smithsonian Institution, and also my uncle, Professor Ernst Mayr, Harvard University.

In addition to the above mentioned, I received other forms of help, for example, references to literature and to specific quotations or illustrations, from Silvio A Bedini, Dr. Bernard S. Finn, Dr. Brooke Hindle, Carlene E. Stephens, David Todd, and Kay Youngflesh of the Smithsonian, Dr. Reinhard Bachmann, Bernard Boissel, Dr. Michael Davidis, Stephan Dietrich, Peter Friess, Barbara Gumbel, and Margaret Nida-Rümelin of the Deutsches Museum, Professor William L. Hine of

York University, Ontario, Canada, Sister Suzanne Kelly, North Easton, Massachusetts, Dr. Svante Lindquist, Royal Institute of Technology, Stockholm, Sweden, Professor Samuel L. Macey, University of Victoria, British Columbia, Canada, Professor Ludolf von Mackensen, Staatliche Kunstsammlungen, Kassel, Professor Nelly Tsouyopoulos, University of Münster, and Dr. Heribert Nobis, Munich.

This work would have been impossible to conceive or complete without the help of numerous libraries, notably those of the Smithsonian Institution and the Deutsches Museum, the Folger Shakespeare Library, and the Library of Congress. I am especially indebted to the staff of the History and Technology Branch of the Smithsonian libraries and to the library staff of the Deutsches Museum, who, with endless patience and endurance, provided me with all required literature. Dr. Paul Forman, Smithsonian Institution, Professor Francis C. Haber, University of Maryland, Professor Thomas P. Hughes, University of Pennsylvania, and two unknown readers of the publisher read the manuscript and gave me many valuable suggestions and comments.

The various drafts of the manuscript were typed into a word processor by Joanna Kofron of Arlington, Virginia, with singular generosity on her own time.

I hope that the merits of the final product will prove the efforts of all these friends and colleagues worthwhile. In any case, all those who have helped have my deep gratitude. To my wife Louise, who also helped in a multitude of ways, the book is dedicated.

Introduction

This book is an essay about the character of technology. It tries to argue two points:

—Technology as a fundamental human activity is intimately related to all other human activities and thus is an integral, indispensable part of all human culture and is not, as one often hears, an alien, inhuman force unleashed upon mankind by some external agent.

—The interactive relationship between technology and all the other manifestations of human life and culture can be *proven,* even interactions as intractable and elusive as that between the political, social, economic, or religious ideas dominant in a given society and contemporary preferences and designs of technological hardware.

The attempt to prove these two assertions takes the form of the analysis and solution of a specific historical problem. This problem is significant and interesting in itself but mainly serves as the vehicle for discussing larger, more general, and more important questions. The specific problem itself is somewhat complex, and so is the strategy employed to solve it and thus to reach the intended larger objective. It seems appropriate, therefore, first to outline carefully our specific problem and the approach proposed for its solution.

Our discussion takes its departure from the following problem. A chronological survey of feedback mechanisms in history has shown that such mechanisms had been known and used in antiquity and had survived through the golden age of Islam, that is, up to the thirteenth century.[1] Feedback devices, however, were not mentioned in Europe, from the Middle Ages through the Baroque. They first reappeared in eighteenth-century Britain, with the Continent lagging significantly behind. This observation, curious in itself, receives added spice from indications that Renaissance Europe was not merely innocently ignorant of the principle of feedback, but that it had deliberately disregarded it. Hero of Alexandria's *Pneumatics* (written probably around A.D. 60) was first printed, in Latin translation, in 1575. The book was eagerly received, as we can tell both from the frequency of its subsequent editions and translations into European vernaculars and from its conspicuous impact upon European technology. Much of its content—syphon arrangements, various automata, turbines, thermoscopes, vacuum devices—was almost instantly absorbed into contemporary theory and practice. Nobody, however, is known to have copied the various feedback devices that form a substantial part of Hero's compendium.[2]

This leads to the double question: Why were feedback devices ignored and rejected in Continental Europe well into the eighteenth century? And why, at the same time, were they cultivated and appreciated in Britain?

Which of these two questions, to begin with, is the controlling one? At first I concentrated on the latter. I had noticed that, in eighteenth-century Britain, the principle of the feedback loop had come into use not only in practical technology but also in abstract arguments, notably in Adam Smith's economic theory. Assuming that this was not coincidental, I tried to establish a connection. I tried to show that the use of the concept in abstract argument had been inspired by practical technology.[3] The attempt failed; in the end, I became convinced that the connection was not direct, but that each phenomenon independently was the result of some unknown earlier cause. This suggested that I take a closer look at the first question: Why was feedback rejected on the Continent?

The question could not be answered in terms of developments in practical technology alone. Granted, the Industrial Revolution in England had stimulated innovation and rewarded inventors at a higher rate than the Continent had done previously. Granted, even, that the Continent so far had little practical need for feedback mechanisms. It hardly had a more practical need, however, for the countless complex clocks and automata that it had been producing for centuries and for all the fanciful and often patently unfeasible inventions that filled the Continental technical literature of the time. Perhaps the answer was to be sought on some other level. Apparently, certain kinds of mechanical inventions were immensely popular on the Continent, while others, including feedback devices, were not. What made the difference?

The external character of the various machines may give us a hint. The purpose of a feedback mechanism is to maintain equilibrium. Its performance is subtle and undramatic: when it performs best, it seems to do nothing. Its hardware is unimpressive, and it is usually just part of some larger machine.

The character of early European technology could not have been more different. Its structures tended to be monumental. Its outstanding inventions were machines that employed large forces and moved conspicuously: wind-and-water wheels, pumps, bellows, cranes and hoists, "*Stangenkünste,*" three-masted sailing ships, siege cannons, and church organs. Such machines, although shaped by the practical needs of the age, exhibited a predilection for dramatic action, a delight in the wielding of raw mechanical power, a pride in the newly attained mastery over nature.

If such were the characteristic machines of the age, then one can understand why feedback devices were ignored. When a class of technology fails both in finding practical applications and in stirring the popular imagination, what is there to save it? The early European indifference to feedback, all this seems to suggest, is best explained in terms of that society's overall attitude toward technology, an attitude that was, in turn, an expression of the people's general outlook, values, and ways of thinking.

Our task, then, will be to reconstruct an instance of interaction between a society's practical technology and its intellectual and spiritual culture. Specifically, we will need to find out: What precisely were the values and thought patterns affecting, and affected by, this interaction? What were the processes and channels in which the interaction took place? How can a historical process like this interaction, which is carried out by its agents largely without conscious awareness and hence without commentary, be documented and reduced to objective proof?

A key to this problem may lie in the spectacular flourishing of a technology that was not only promoted and supported with unparalleled enthusiasm by all levels of society but that also had a rather meager justification in practical utility: we must take a close look at that amazing production of clocks and automata in late medieval and Renaissance Europe. To get an idea of the values, aspirations, and thought patterns of the time and to see how these found expression in technology, it will not be enough to analyze the mechanisms of the clocks, to appreciate their mechanical and artistic perfection, or to comprehend the unprecedented quantities in which they were produced. We must also consider the extent to which these clocks and automata affected everyday life in general. For this, we will need contemporary comments and statements about clocks, especially testimony that is not merely descriptive but which, implicitly or outright, expresses a valuation.

This study will proceed as follows: in the first part, it will explore attitudes toward technology by analyzing a large number of metaphorical references to clocks and automata in early modern European literature. This will yield empirical evidence of differences in attitudes and values between the Continent and Britain that go far in accounting for the contrasting receptions given to the principle of feedback. We will perceive two divergent developments: a growing commitment on the Continent to the value of authority, and in England, to its antithesis, liberty. The quintessential symbol for authority, of course, was the clock.

The second part will investigate, by the same method, the leading metaphors of liberalism, the notions of equilibrium and balance, and their role in European literature. Concurrent with the rise of liberalism in practical politics, the metaphor of the "balance" reached a degree of prominence in English literature that was never approached in the literatures of the Continent. The simple metaphor gradually evolved into an abstract concept—known under such names as self-balance, self-regulation, or dynamic equilibrium—and finally into the scheme of action that underlies the various basic mechanisms of liberalism—the "checks and balances" of constitutional government and the "supply-and-demand" mechanism of the free market. This scheme, however, is nothing other than the principle of feedback.

Finally, after this introductory outline of the book's themes and approaches, a reminder is in order. On the first level, the book deals with the intellectual history of clocks and feedback control and with the relationships between the mechanical clock and an authoritarian conception of order and feedback control and a liberal concept of order. The intriguing similarities, differences, and interactions of those two relationships are also considered. But on a higher level, these discussions are only instruments for another purpose; they serve to demonstrate, by examples and tangible evidence, that technology is a cause as well as a consequence of the values and norms of the society in which it flourishes—technology is both a social force and a social product.

I Authoritarian Systems

1

The Mechanical Clock, Its Makers and Users

The Middle Ages in Christian Europe were an epoch of extraordinary technological creativity. The period began with an agricultural revolution that allowed feeding a rapidly growing population and, incidentally, restructured society; climaxed with the tradition-defying engineering of the Gothic cathedrals; and ended with the introduction of printing by movable type. Its innovations included, to name only a few, the large-scale adoption of water mills for a wide range of purposes, the introduction of the windmill, nautical instruments (compass and sand glass) and techniques of ship design that made possible the navigation of the open seas, eyeglasses, gunpowder and fire arms, the spinning wheel, and cast iron. Historians explain this outburst of creativity through a happy confluence of factors: the combined heritage of techniques and skills from ancient Greece and Rome, from Islam, and from the Celtic and Germanic tribes north of the Alps; the challenges of a northern climate; the high value placed on practical activity and manual labor by the Catholic Church; and the spread of a new attitude toward nature which insisted that man was nature's master and technology the instrument of his dominion.[1]

The mechanical clock appeared on the scene in the latter half of this fertile period, a little before A.D. 1300. Its inventor is unknown, and so is its place of origin, which may lie anywhere between northern Italy and England.[2] The rapid acceptance of the new clock indicates that it met a definite demand. This demand was strongest in the tightly organized communities of monasteries and towns.[3] The monasteries needed to announce, in a regular, predictable manner, the hours of prayer at day and night around which their activities were structured. The towns, usually self-governing, with people's livelihoods derived from commerce and industry, found it more and more necessary to order civic life according to stated schedules. Traditionally, the hours both of religious services and of town activities, such as curfew, market time, or the opening and closing of the gates, had been struck by hand on special bells. If anyone needed a reliable clock, then it was the tower warden who had the job of striking the hours. In the past he had had to make do with traditional devices, such as sundials in daytime and water clocks at night, the limitations of which were even more painfully evident in the cloudy skies, long nights, and cold winters of northern Europe than they had been earlier in Mediterranean latitudes. The tower warden was probably the first user of the new mechanical clock and was also the first to be replaced by it once this clock had advanced enough to strike the hours automatically.

The new invention owed a great deal to traditional horology and to inherited technology. The basic conventions of time measurement had been shaped by the oldest timekeeper, the sundial, which had been in use in early Egypt and had become highly refined and widely popular in Greek and Roman times.[4] It gave rise to the custom of regarding day and night as separate entities, demarcated by sunrise and sunset, and to the convention of dividing day and night each into twelve parts, which varied in length with the seasons—temporal or seasonal hours. When the mechanical clock was invented, the sundial was as popular as ever, and it remained so, due to its simplicity and inherent reliability, for several centuries. Well into the nineteenth century it served as the basic time standard by which clocks and watches were set. Its shortcomings had been recognized, however, even in antiquity. Not only was its usefulness restricted by a dependence on sunshine; it also lacked the capability of generating output power, that is, of advancing hour hands, striking bells, or powering automata displays.

The water clock had its strengths precisely where the sundial was weakest. Simple water clocks are recorded in ancient Egypt and Babylon in the late second millennium B.C.[5] In classical Rome, as reported by Vitruvius and Hero of Alexandria, water clocks were built in great variety and sophistication; occasionally they took the form of large water-powered monumental clocks with many subsidiary indications and automata displays, installed conspicuously in public buildings for the entertainment of urban populations. All water clocks had in common a water vessel with a slowly changing level that, tracked by a float and transmitted through levers and linkages, served not only as a timing element but also as a source of power that could drive surprisingly large mechanisms. Water clocks were thus closely related to the highly developed ancient technology of automata: the steady sequence of hourly indications offered itself as a convenient way of programing automata performances. This was perhaps, in the long run, the most important contribution of the ancient water clock: it taught how to regulate machinery over long spans of time by means of program control.[6]

There was yet a third strand of technological heritage that was essential for the invention of the mechanical clock, namely, the technology of toothed wheels and the complex gear trains composed of them. By the end of the last century B.C., as the well-known astronomical computing device from Antikythera proves, these techniques had advanced to an amazing degree of refinement.[7] Gear trains were used primarily in instruments indicating the motions of the planets relative to zodiac and earth or to express the relationships between the various astronomical and calendarial cycles. That this ancient technology survived with notable continuity is evident from actual surviving specimens of geared mechanisms as well as from pictures, direct descriptions, and literary references from the late Antiquity, Islam, and early medieval Europe.[8]

On the eve of the invention of the mechanical clock, these three threads of ancient technological tradition—sundials, water clocks, and geared indicating and computing instruments—were vigorously alive, separately and in various combinations. By the thirteenth century, the search for a better timekeeper had become energetic; this was expressed in reports of new designs that had finally broken out of the traditional mold and was also stated in direct testimony.[9] In 1271 an English cleric teaching at one of the French universities described the current state of horology.[10] To him there was no longer any doubt that the clock of the future would be mechanical and weight driven, but one problem remained: lacking was a device (which we now refer to as the *escapement*) to retard the accelerating motion of a falling weight into a steady motion of strictly constant velocity. This key element, however, and with it the successful mechanical clock itself, was invented within the following decade or so.

The *verge-and-foliot* escapement was an invention of radical originality (see figure 1–1). The complexity of its functioning and the ingenuity with which it had been devised far surpassed all previous mechanical inventions. Its characteristic feature, the carefully tuned dynamic interaction of several matching parts, is almost impossible to describe in words or to illustrate in a two-dimensional picture; to fully comprehend it, one should handle a working specimen.

The exceptional character of this invention makes our ignorance about its origins all the more tantalizing. It has been suggested that the verge-and-foliot escapement was an import from the East, but most historians of horology believe it to be an indigenous European invention.[11] The extreme fertility for invention of medieval Europeans in general, the rapidity with which the mechanical clock spread, and the facility with which its basic mechanism was supplemented by other, similarly complex, inventions seem to justify the latter assumption. Thus, a little before A.D. 1300, the mechanical clock came into being. Its

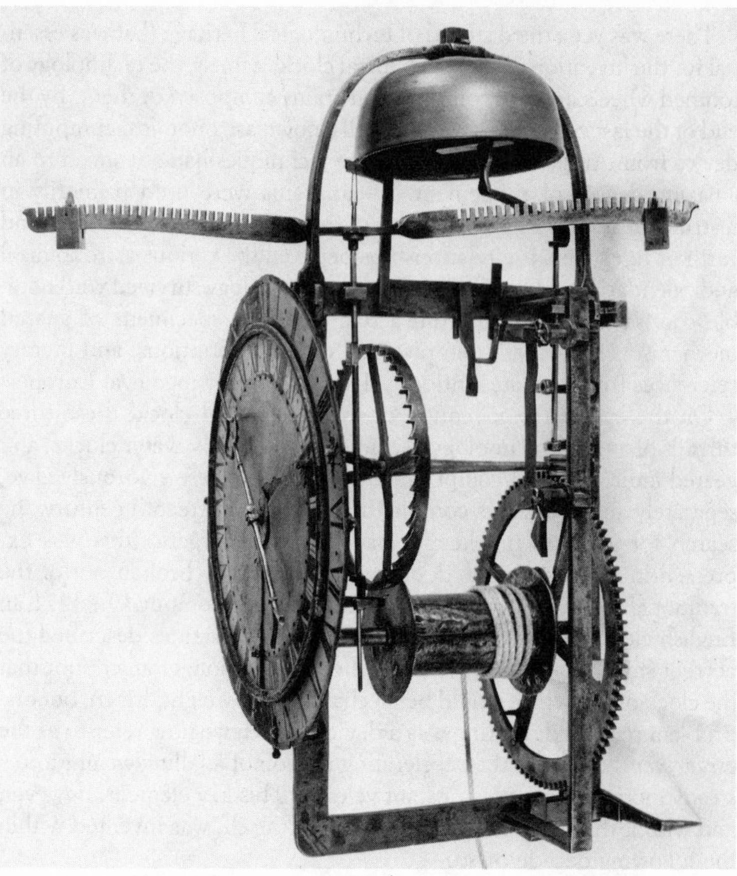

Figure 1–1 Nuremberg fifteenth-century wall clock; a typical weight-driven Gothic clock, capable of indicating the hour and of ringing an alarm bell at a preset time (Germanisches Nationalmuseum, Nuremberg).

technical development during the following several centuries proceeded in three phases, initiated, respectively, by the basic invention circa 1300, by the introduction of the spring drive circa 1500, and by the invention of the pendulum clock, 1657.

The clocks of the first phase, from roughly 1300 to 1500, with few exceptions, were weight driven. Most of them were large tower clocks; small domestic clocks were comparatively rare. These clocks were made by wandering clockmakers, by practitioners of older crafts such as black- and locksmiths, and by scholarly amateurs such as theologians, astronomers, and medical doctors. Despite such limitations in productive capacity, the clock spread over Europe with remarkable speed.

With production concentrating on large tower clocks for installation in churches, town halls, and city gates, the benefits of the new timekeeper were brought quickly to the largest possible number of people. By the middle of the fourteenth century, weight-driven mechanical clocks indicating and striking the hours, and often capable of other more complex displays, were installed in full view of the public in all major cities and were rapidly spreading to the smaller towns.

Such rapid introduction of the clock into broad popular use would have been impossible without an equally rapid maturing of the mechanism in the technical sense. Basic auxiliary mechanisms like alarms, striking trains, and carillons had been added shortly after the original invention. More complex machinery, for example, as used for astronomical indication or for automata performances, appeared on the scene almost at the same time, probably adopted directly from traditional technologies. Within half a century from its invention, the mechanical clock had advanced enough to make possible such outstanding achievements as the astronomical clock of Richard of Wallingford, the abbot of Saint Albans in Hertfordshire (ca. 1330); the monumental clock of Strasbourg Cathedral (1352) (figure 1–2); and Giovanni de'

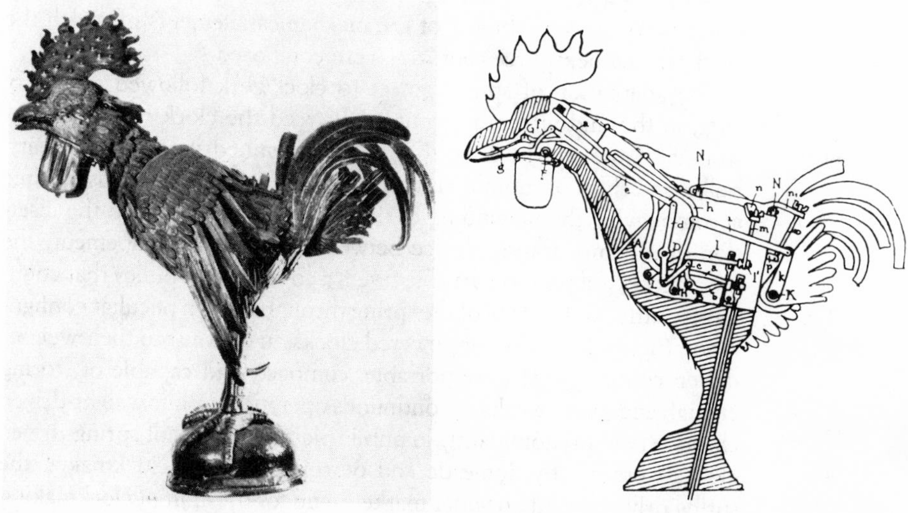

Figure 1–2 (A and B) The rooster automaton of the first Strasbourg monumental clock, 1354; measuring 122 cm high, it is the only surviving part of this clock. It was mounted near the top, as it was again on the clock's 1574 successor (see figure 1–7) and was capable of performing an elaborate program of automatic motions and crowing (Musée des Arts décoratifs, Château des Rohan, Strasbourg).

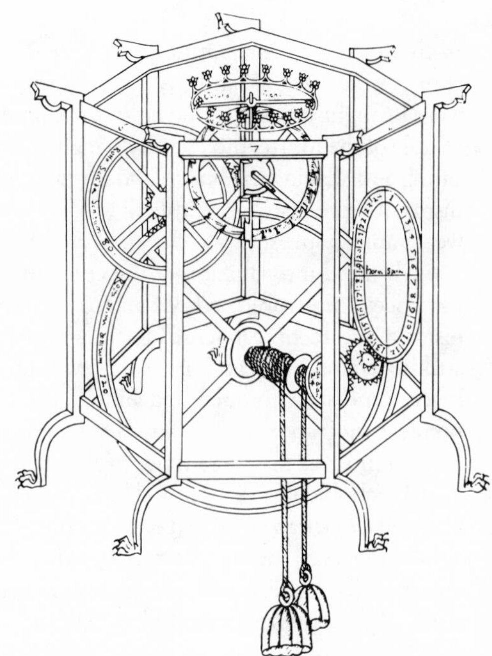

Figure 1–3 The planetary clock of Giovanni de' Dondi, Padua, between 1348 and 1364; the inventor's own sketch shows the lower frame with the going train of the clock. Not shown is the upper frame with the epicycloidal gear drives for the movements of the seven planets.

Dondi's *astrarium* in Padua (1348–64) (figure 1–3).[12] These three early masterworks included most of the mechanical elements of which the clocks of the next three centuries were composed.

The adaptation of spring power to clockwork followed a century later, in the mid-fifteenth century.[13] It freed the clock from the constraints of the weight-drive that had condemned it to be stationary, bulky, and short of running time. The crucial problem, the loss of force accompanying the unwinding of the coiled spring, required the insertion of a compensating device between spring and escapement; the most successful solution was the fusee, a cone-shaped pulley that counteracted the nonlinearity of the spring through its own peculiar configuration (figure 1–4). Spring-powered clocks, in contrast to their weight-driven counterparts, were portable, compact, and capable of storing enough energy for weeks of continuous operation. While weight-driven clocks served predominantly in public places, the uses of spring-driven clocks were mostly domestic and personal. For the clockmaker, the spring drive opened up wider markets, and for the craft of clockmaking it initiated an era of professionalization.

Early in the sixteenth century, clocks and watches were produced systematically and in quantity. Their artistic and technical quality rose dramatically. Responsible for this combined jump in quantity and quality was a change in the social constitution of the craft: clockmaking had

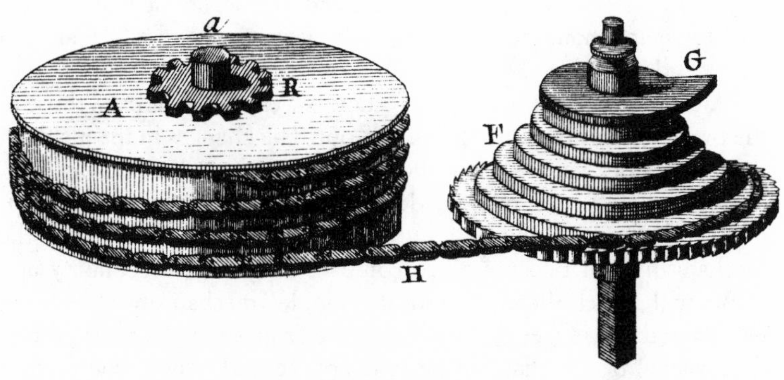

Figure 1–4 Fusee (right) and spring barrel (left).

become the specialty of full-time professionals. Clockmakers began to form craft associations, to concentrate in certain localities, and to produce not just for local needs but for a wider national, and indeed, international market. From about 1500 to 1650, clockmaking flourished most in German-speaking central Europe; the outstanding centers of the craft were some south-German "Free Imperial Cities," such as Strasbourg, Ulm, Nuremberg, and, above all, Augsburg.[14] These cities owed their preeminence to a combination of factors. The region was traditionally famous for the skill of its metalworkers. Recent innovations in metal mining and processing, which made Germany the international leader in these fields, provided superior materials at low cost. City republics that depended on trade and craft production for their prosperity provided a favorable environment for clockmaking in general; they offered protection and support to the communities of clock workers as a whole, especially in granting the right to form separate guilds.

These circumstances permitted an impressive output. Augsburg, the leader in clock production and also Germany's largest city at that time, licensed during the period of 1550 to 1650 a total of at least 182 master clockmakers; accordingly, at any given moment during this period, it must have had some fifty to seventy independent shops, most of them staffed, in addition to the master, with at least one journeyman and one apprentice each. Productivity was enhanced by much standardization and specialization and by extensive use of prefabricated materials and parts. We get a sense of the quantities involved once we realize that a shop could turn out a simple clock in about two to three weeks, or fifteen to twenty-five per year. The time allowed a journeyman for his masterpiece (figure 1–5), however, was half a year. At any rate, the clocks produced in Germany over the period from about 1550 to 1650 counted many thousands, a surprisingly large portion of which are still extant today.

Clocks made under these circumstances were different from those of the previous century, both inside and out. The changes in the inside of the clockworks were subtle rather than dramatic and are best characterized as a shift from homemade ingenuity to sophisticated engineering. While the capabilities of a given movement increased greatly, the mechanism itself became more compact in size. The economy and clarity with which these enormously complex mechanisms were fitted into limited spaces (see figure 1–6) showed a mastery of rational, three-dimensional design that can be fully appreciated only in comparison with the best work of the nineteenth and twentieth centuries.

More dramatic were the changes in external form. Renaissance clocks were self-sufficient works of art. While Gothic clocks had open cages with all of the gears in full view, Renaissance clocks were housed in ornately sculptured cases that satisfied the standards of architecture and the decorative arts of that period. They were created fully in three dimensions and were freestanding, with indications on all four sides—totally unlike the clocks of today with their single dial and one-dimensional manner of display. Horological utility was often subordinated to artistic ambition, and clocks were housed in sculptures of mythological animals, crucifixions, or vases. Timekeeping, it seems, was no longer a prime concern. Among the profusion of indications of a typical Renaissance clock, the hour dial is often hard to discover. Minute hands—while technically quite feasible—were uncommon, and second hands appeared on only a few rare experimental clocks built for astronomical observatories.

These clocks were not very accurate (to within a few minutes per day, at best), and they were too complex to be reliable. Their objective, it seems, was universality; they attempted to mirror the whole human experience. It could be claimed, as did Conrad Dasypodius, the scholarly designer of the second Strasbourg Cathedral clock (1572), that clocks comprised the entire quadrivium, namely, the major four—astronomy, geometry, arithmetic, music—of the seven liberal arts.[15] Apart from timekeeping, their principal capabilities were astronomical prediction, musical performance, and the imitation of life. These capabilities were exercised on any scale and in almost any combination; the great monumental clocks of some cathedrals—Strasbourg was the best known—comprised all four. Astronomy, nevertheless, ranked highest; a few astronomical indications were offered on almost every clock, and fully equipped "astronomical clocks" were remarkably common (see figure 1–7). These astronomical indications were based on the

Figure 1–5 Table clock, presumably a masterpiece, maker unknown, Augsburg, 1600. The clock's overall height is 52 cm, and it has three separate spring-driven gear trains (for the going train and the hour- and quarter-hour-striking trains). It presents horological and astronomical indications on all four sides, including dials for the minutes and hours and for the length of days; an astrolabe showing the sun's progress through the zodiac; the aspects, age, and phases of the moon; and calendar disks for saint's days; the dominical letters, the dates of Easter for a century, and so forth. The clock was also an alarm clock and performed automatic displays every hour (Bayerisches Nationalmuseum, Munich).

Figure 1–6 (A and B) Table clock by Steffen Brenner, Copenhagen, 1558; the clock, with separate spring-driven going, hour-striking, and quarter-hour-striking trains and an alarm mechanism, presented a multitude of indications not only on all four sides but also on top. It is 21 cm in height and has a twenty-four-hour dial with astrolabe, including a dragon hand with rotating moon image and a sun hand; a dial for the age and phases of the moon; a diagram for the lunar aspects; and dials for the length of day and night and for the days of the week; a calendar dial; and so forth (St. Annen Museum, Lübeck).

old geocentric world system, and their significance was largely astrological. Some of these indications—for example, the motions of sun and moon relative to the zodiac, the length of day and night in the course of the year, changes of the sun's velocity during its course around the ecliptic, the irregular motions of planets (although these were only seldom shown)—occasioned engineering feats of high distinction.

In a technological sense, the golden age of German clockmaking was not an unusually innovative era but rather a time of consolidation and refinement. As an economic force, German clockmaking had spent itself by the mid-seventeenth century; weakened externally by the economic and social tolls of the Thirty Years' War and internally by the restrictive conservatism of the craft guilds, it quietly slipped into the backwater of horological advances made by its Western neighbors.

In the third quarter of the seventeenth century, a series of inventions made in Holland and England set into motion what has been called a horological revolution.[16] Christian Huygens's invention of the pen-

Figure 1–7 The second astronomical clock of Strasbourg Cathedral at its completion in 1574; it was designed by Conrad Dasypodius and built by Isaak and Josias Habrecht. Woodcut by Tobias Stimmer, 1574.

dulum clock in 1657 (anticipated by Galileo) achieved a sensational increase in accuracy. The invention of the balance spring extended the same benefit to the pocket watch. Kinematically refined escapements and temperature-compensated pendulums brought further improvements. By the middle of the eighteenth century, it became possible for the first time, both in England and France, to build timepieces—chronometers—that possessed the extremely high accuracy necessary for determining longitude at sea.

The horological revolution was not only the result of a few crucial inventions; it also reflected a new set of economic and social circumstances. Its home was in Holland, England, and France, which were powerful, prosperous nation-states with central governments that consciously cultivated commerce and industry. Both England and France had royal academies, by means of which the state tried to direct its best scientific talent to the promotion of technology and industry. Characteristically, outstanding contributors—Galileo, Christian Huygens, and Robert Hooke—were no longer craftsmen but theory-oriented scientists associated with national academies. The leading national capitals, London and Paris, were not only seats of academies but also fast growing concentrations of commerce and industry with huge urban populations, a considerable proportion of which was a prosperous and growing bourgeoisie. These cities became the new capitals of clockmaking.

Increased accuracy was not the only achievement of the horological revolution; clocks also became simpler and less expensive. Their external appearance changed radically. With only a single dial showing the hours and, as a matter of course, minutes, the new clocks were housed in functional wooden cases that could be austere to the point of literally being "black boxes" (figure 1–8). The machine that had mirrored the universe turned into an unpretentious precision instrument with only one function, timekeeping. The watches, previously plump and ornate, to be worn proudly around the neck, became plain and flat and disappeared into the vest pocket. In 1776, Adam Smith could state that "a better movement of a watch, than about the middle of the last century could have been bought for twenty pounds, may now perhaps be had for twenty shillings."[17] The horological revolution was complete, and the character of the mechanical clock has not changed much since then.

This survey, however brief, of the internal-technical development of the mechanical clock makes it clear enough that it was a machine unlike any other. The impression will be reinforced by considering how these mechanical clocks were used. On the most elementary level, clocks were timekeepers. But, even by the standard of practical utility, they were not strictly comparable to machines such as a mill, a loom, or a hoist. Time measurement was not an elementary human need like food, clothing, or shelter; if it could be called a practical need at all, it was a derivative one, peculiar to a highly advanced civilization. In medieval Europe, the strongest need for a reliable timekeeper had arisen in the management of organized communities. The mechanical clock had been invented, it seems, in response to a need of monasteries and cities. And, as this need was being satisfied, life in these communities was transformed. The process began when, during the fourteenth century, large tower clocks were installed in the churches, town halls and city gates of the major towns. With the automatic tolling of the hour bells, which were audible for miles, it became possible to subordinate the public life of the whole town to a common schedule. The activities of the citizens could be coordinated in more and more tightly interlocking patterns. In designs of utopian communities, it became customary to place a clock, in dominant position, in the center of the town.[18]

The public welcomed the new timekeeper with almost unanimous enthusiasm. The prevailing sentiments were expressed by one author who praised a well-kept clock as the sign of a well-run community and by another who called clocks the very criterion of civilized existence.[19]

There were only a few literary protests against the clock's expanding domination: Dafydd ap Gwilym, a Welsh bard (fl. 1340–70), exclaimed with passion:

> Woe to the black-faced clock which awoke me on the ditch side. A curse on its head and tongue, its two ropes and its wheels, its weights, heavy balls, its yards and its hammer, its ducks which think it day and its unquiet mills. Uncivil clock like the foolish tapping of a tipsy cobbler, a blasphemy on its face.[20]

Similar sentiments are expressed in Rabelais's *Gargantua,* where one of the story's heroes, the Monk, boasts that he will never be subjected to a schedule: "The hours are made for man, not man for the hours." Approving of the Monk's outlook, Gargantua built him an abbey that was decreed to contain "neither clock nor dial."[21] Voices of hostility to the clock, however, such as Dafydd's or Rabelais's, were rare.

As the clock extended its dominion, the concept *hour* acquired new meaning. Life had traditionally been ordered according to the cycles of earth, sun, and moon, with the sundial as the principal instrument of time measurement. Day and night were independent quantities, each divided separately into "temporal hours" that varied with the seasons and with latitude. The mechanical clock, however, oblivious of the cosmic rhythms, favored units of time that were rigidly constant. It forced the introduction of "equinoctial hours," which integrated day

Figure 1–8 After the Horological Revolution: Standard bracket clocks were equipped with highly accurate pendulum movements offering only a minimum of indications and were housed in austerely simple cases befitting their Protestant places

and night into a uniform twenty-four-hour period, independent of sunrise and sunset.

As the towns, with the ready approval of their populations, ordered public life according to the clock, individuals also welcomed the clock into their private lives. Princes and their courts led the way. Courts, with their complex ceremonials and their many-layered staff hierarchies, demanded punctuality from everyone; punctuality became the "courtesy of kings." Punctuality and the habit of disciplining one's daily life

of origin. (A) Bracket clock by Johannes van Tegelberg, The Hague, circa 1670 (National Museum of American History, Washington, D.C.), (B) Bracket clock No. 99 by Thomas Tompion, London, circa 1690 (private collection).

were methodically inculcated into young princes: their daily regimens were laid out on detailed schedules. The clock became an attribute of nobility. It was fashionable, from the sixteenth well into the nineteenth century, that clocks or watches would be featured prominently on portraits of dignitaries (see figures 1–9 through 1–11). The clock's metamorphosis from a timekeeper to a symbol of aristocratic rank reached comical extremes when, in the seventeenth century, would-be gentle-

Figure 1–9 Portrait of a young man, presumably Karl von Burgau, son of Archduke Ferdinand of Tirol, painter unknown, circa 1590 (Kunsthistorisches Museum, Vienna).

Figure 1–10 Portrait of Cardinal Richelieu by Philippe de Champaigne, 1622 (Le Musée Condé, Chantilly).

Figure 1–11 Portrait of Don Justino de Neve by Bartolomé Esteban Murillo, circa 1650 (National Gallery, London).

men were discovered wearing around their necks watches that were empty.[22]

An important aspect of the appeal of the mechanical clock, which had nothing to do with timekeeping at all, was rooted in the magic of self-moving machines. From the beginning, clocks were equipped with multiple capabilities. Of the three most famous clocks of the fourteenth century, those of Richard of Wallingford, Giovanni de' Dondi, and Strasbourg Cathedral, the first two were primarily astronomical automata with hour dials as incidental by-products, while the last, apart from its time-telling and astronomical capabilities, could play a mechanical carillon and activate automata figures. This tendency of emphasizing the automata capabilities of clocks while downplaying their time-telling capability culminated in the sixteenth and seventeenth centuries. Automata performances not only became standard requirements for the larger tower clocks; they were also realized on various smaller scales, down to that of pocket watches. Clocks and automata, in short, tended to be very much the same thing (figure 1–12). Contemporaries recognized this clearly: in dictionaries from the sixteenth to eighteenth centuries, the automaton (a machine that moved independently on the strength of both a power supply and a plan of action, or program, of its own—see figures 1–13 and 1–14) was the higher, general category; the clock was merely a particular variety of automaton.[23]

In practice, automata tended to fall into two categories: *astronomical clocks* that attempted to reproduce, as comprehensively and precisely as possible, the motions of the heavenly bodies and *figure clocks* that imitated, often accompanied by mechanical music, the actions of living beings.

The astronomical clocks served in medical and horoscopical astrology by indicating the positions of the planets in the zodiac (figure 1–15);[24] they provided calendarial information ranging from the times of sunrise and sunset, the age and phases of the moon, and the saints' days to data for the prediction of eclipses or Easter. They were also of some use to observing astronomers as continually up-to-date reference devices.[25] How meaningful these astronomical indications were to the casual viewer is hard to judge. One suspects that some of those ingenious and beautifully crafted indications served primarily to be shown off to guests and to enliven conversation. We must recall, however, that people of the Renaissance were far more familiar with and interested in the phenomena of the nightly sky than we are today.

An element of frivolity was even more evident in the use of figure clocks (figure 1–16). These automata found employment in the table decorations, drinking games, festive receptions, and parades of princely courts.[26] Common people, too, had opportunities to see automata in

Figure 1–12 Astronomical clock with automata displays, modeled after the Strasbourg monumental clock, by Isaak Habrecht, Strasbourg, 1594 (Rosenborg Castle, Copenhagen).

Figure 1–13 The program-controlled mechanical pipe organ of the table carriage "Minerva" (for overall view, see figure 1–16), presumably by Achilles Langenbucher, Augsburg, circa 1630. Note the large, pin-studded program drum (Kunsthistorisches Museum, Vienna).

Figure 1–14 Tower clock, Marburg, seventeenth century; note the lockplate, front center—the number of the hourly strikes is programmed into the distances between notches at the periphery (Universitätsmuseum, Marburg).

Figure 1–15 Planetary clock by Eberhard Baldewein, Marburg, 1558–62; the spring-driven clock presents on its four sides the epicyclic movements of the planets and an astrolabe. On top it features a celestial globe, where the sun is automatically moved along the ecliptic (Staatliche Kunstsammlungen, Astronomisch-Physikalisches Kabinett, Kassel).

action, be it on the monumental clocks installed in public buildings or at fairs and carnivals, displayed by itinerant showmen.[27] The viewer of the sixteenth or seventeenth century was probably vastly more impressed by the experience than we would be today, for he was unfamiliar with such apparatus and was purposely kept at a distance from the performing automata, with the circumstances of the display carefully stage-managed to produce a mood of mystery and magic. Being accustomed to accounting for unexplained phenomena in supernatural and magical terms, he doubtless responded with strong emotions and remembered the experience vividly.

Medieval and Renaissance society's fascination with automata had roots in classical antiquity. The notion of building automata in the strict sense—artificial living beings—had fascinated the ancient Greeks. An-

Figure 1–16 Three automaton table carriages: (left) "Amor," by Andreas Stahel, Augsburg, early seventeenth century (Badisches Landesmuseum, Karlsruhe); (center) "Diana," South Germany, early seventeenth century (Museo Poldi Pezzoli, Milan); (right) "Minerva," Augsburg, circa 1625 (Kunsthistorisches Museum, Vienna). With the help of spring-powered clockwork, these carriages were capable of driving under their own power, indicating the time, playing mechanical music, and performing a variety of automaton displays.

cient legends of automata built by mythical inventors began to be collected and repeated in the late Middle Ages and achieved great popularity in the sixteenth and seventeenth centuries.[28] To the tales of the automata of Daedalus and Vulcan, the wooden dove of Archytas, and the water-powered world machine of Archimedes, to name only a few, Renaissance authors would add, in the same breath, reports of more recently built automata, for example, the life-size mechanical woman built by Albertus Magnus which Thomas Aquinas, his disciple, had smashed as a work of the devil; the lion built by Leonardo da Vinci in Milan for the reception of the French king; and the iron fly and eagle of Regiomontanus in Nuremberg.

Compilations of such automata legends were repeated in dozens of books of the sixteenth and seventeenth centuries. The authors were curiously uncritical; they were not interested in distinguishing improbable tales of demons and sorcery from documented feats of engineering.

Their attention, instead, was on a subject that was forbidden and yet irresistable—magic. The automata legends provided historical legitimation for a new form of magic that had the powers of "black" magic but that required no help from sinister forces: the technology of automata was characterized variously as "mechanical," "mathematical," or "natural" magic. Clocks and automata thus became mediators between the worlds of magic and rationality; demonstrating how magical effects could be achieved by clever machinery of metal and wood, they changed the mental habits of their viewers. Gradually people learned how to account for unfamiliar phenomena not by resorting to the supernatural but by identifying concrete, observable causes; they learned how to connect causes and effects into extensive, characteristically linear cause-and-effect chains. Clocks, in short, helped to teach Europeans how to think "mechanically."

In sum, the purely practical uses of the clock, then, were the following:

—As a timekeeper, it made possible the tighter coordination of the activities of increasingly complex communities.

—As a mechanical representation of the known universe, it was a repository of information, a teaching tool, an astrological accessory, and perhaps, to some modest extent, an instrument for the scientific astronomer.

—As an automaton, its applications ranged from entertainment of drinking partners to the production of effects of magic before uninitiated audiences.

Such a list of the practical uses of early clocks, diverse as they were, is misleading, however, because the motives from which clocks were built and acquired were far more complex. On the lowest level, many clocks were acquired not to be used but merely to be owned. Their principal function was to give pleasure to the owner and to be shown off to friends and guests. On the highest level, by contrast, every generation exerted itself to build certain superlative clocks intended to surpass everything that was known to date. Efforts of this kind, especially before the horological revolution—the Strasbourg Cathedral clock of 1572 is a typical example—had a wide array of capabilities but no concretely definable practical function. They can only be viewed as exhibitions of mechanical genius and as celebrations of man's mastery over nature.

Were one, in conclusion, to summarize the significance of the mechanical clock in early modern Europe, one might draw attention to the following features:

—The mechanical clock was a creative achievement of high intellectual rank. Nothing of comparable ingenuity had ever been invented before, and up to the advent of the steam engine, it remained Europe's intellectually most demanding mechanism.

—Early modern Europeans made extraordinary commitments to this mechanism. For centuries, clockmaking had been the preferred field of the best technological talent. And society at large supported clockmaking with extraordinary material resources.

—The numbers in which clocks were produced, especially after the mid-sixteenth century, are disproportionately large in comparison with all other contemporary machinery.

—No other historical machine has been preserved through the centuries with the same untiring devotion. Even today, our private and public collections contain thousands of clocks from the Renaissance.

Notwithstanding its distinction as a milestone in the history of mechanical inventions, what made the clock unique was the unprecedented veneration that it commanded in the society of that day. In order to understand this fascination, it is not sufficient to know the conventional technological, economic, and social history of the mechanism. What is necessary is to find out how contemporaries not only spoke but also thought and felt about clocks. What we need is access to their un- and subconscious minds.

2

The Rise of the Clock Metaphor

Clock Metaphors as Evidence

We cannot cross-examine people from distant centuries about their passion for clocks. Besides, they might have made poor witnesses. How could they have articulated the unconscious and subconscious roots of their fascination with this machine? Fortunately, they left us direct testimony of another kind that is probably much better. There was a widespread habit of referring to clocks in writings on almost any subject in the form of comparisons and metaphors.[1] Clock metaphors, then, will be the empirical data upon which the following chapters are based.[2] Representative early modern European literature will be sifted for clock metaphors of every type. The resulting harvest will be analyzed and then classified pragmatically into whatever categories offer themselves. Chapter 2 deals collectively with the early period when clock metaphors were not yet too frequent or specialized. After the early seventeenth century the metaphor became so popular, especially in natural philosophy and political thought, as to require separate treatment in chapters 3 and 4. Chapter 5, finally, will evaluate the accumulated empirical evidence in its totality.

As testimony about clocks, metaphors have the advantage of being unselfconscious: when an author introduced a clock metaphor in support of some argument, it was on this argument that he concentrated his attention, not on the metaphor or the clock. Through such metaphors, therefore, he is more likely to reveal his true feelings about clocks than through any conscious statements.

How frequent, relatively and absolutely, were clock metaphors in early modern European literature, and how reliable is our knowledge about such frequency? A reliable assessment would have to be based on rigorously statistical methods: to begin with, some defined body of testimony, say, all written records of Europe from 1300 to 1800, would have to be sifted exhaustively for clock metaphors, and the significance of the findings would have to be weighed in comparison with the frequency of other popular metaphors. For obvious reasons, such an approach would have been impracticable in the present circumstances. The only realistic alternative was to collect as many specimens of metaphors as possible with the available means, that is, to concentrate on the published literature, by favoring, for reasons of accessibility, publications in English, French, and German, and to forego all claims of exhaustiveness. In the materials thus gathered, one can discern the following quantitative patterns:

—In the fourteenth, fifteenth, and sixteenth centuries, the clock metaphor was not used with conspicuous frequency. Notable, however, was the tone of emphatic approval in which it was normally invoked. In later centuries, when the metaphor became much more frequent, this tone of approval did not change. There were, to be sure, voices that questioned and on occasion rejected the value of the clock, but they were extremely rare except in England. England, as will be shown, played a role in the story that was altogether exceptional.

—In the seventeenth and eighteenth centuries, the clock metaphor became strikingly frequent, more frequent, probably, than any other. Such frequency signifies more than broad public approval of the clock. It tells us something about that society itself. The clock metaphor thus becomes a means of access to the collective mentality.

What were the links between the metaphor and the "collective mentality?" We must recall that metaphors are testimony on behalf of those who wrote them and those who read them. Quite generally, the study of the metaphor yields evidence that is immediately valid only with respect to those who could read and write. Such evidence concerns a rather small portion of society, although one that doubtless grew between 1300 and 1700; it concerns predominantly the ruling classes and the aristocracy, the members of the learned professions, and those who lived in towns. Directly, the metaphor tells us little about the thinking and feeling of that majority of people who were illiterate, without access to literature, or removed from town life. Nevertheless, although the clock metaphor spoke for only a fraction of the whole society, this was an important fraction: it spoke for a minority that governed the country, controlled its economy, and, in the broadest sense, shaped its culture. This group was very nearly identical, of course, with the minority of those who had any familiarity with actual clocks, as owners, as acquaintances of owners, or as town dwellers living within earshot of a tower clock. While the clock metaphors spoke for only a minority, one may assume that it represented that minority fully. Clock metaphors can be found in the writings of countless representative as well as leading figures, and dissent was rare.

If one compares the genealogy of metaphors with those of concepts and ideas, there is a basic difference. Concepts like *trinity* or *momentum* live in history as subjects of continuing discourse. Anyone wishing to participate in this discourse is expected to be knowledgeable about the concept's previous history and to express himself in the terminology of the preceding discussion, no matter whether he wishes to approve of the concept or criticize it. Such constraints on behalf of focus, cohesion, and homogeneity are notably lacking in the use of metaphors. Metaphors, by their nature, are not the *subjects* of the discourse but only auxiliary devices adduced for emphasis and illustration, upon which neither author nor audience will concentrate much attention. Often enough, authors who were in utter disagreement on matters of substance have used the same metaphors. There is no incentive that compels authors to know or respect previous uses of the metaphor; indeed, metaphors are more effective when novel and fresh. Apart from literary custom, the only forces keeping a metaphor focused on a given theme are its inner logic and the suggestive power of its imagery. Accordingly, in evaluating a large collection of clock metaphors, it will be wise not to expect any neat convergence upon a single well-defined theme.

Late-Gothic Beginnings

Some of the earliest clock metaphors come from a distinguished source, from the *Divine Comedy* of Dante Alighieri (1265–1321). That it was the mechanical clock that Dante had in mind and not the water clock, as some have claimed,[3] is indicated both by chronology and by internal evidence. Dante wrote the *Paradiso,* his poem's last canticle, in the last years of his life, three or four decades after the clock's invention. By that time Dante must have seen at least the clock of San Eustorgio, installed in 1307 in Milan, a town in which he is known to have stayed since that date.[4]

There are two significant passages. In the first, Dante, guided by Beatrice, has entered the fourth heavenly sphere, the Heaven of the Sun. A group of spirits, in bright light, form a ring around him, dance, and then stop. One of them, introducing himself as the soul of Thomas Aquinas, identifies the others as the souls of the twelve principal teachers of Christianity:

Then as a horologe [clock] which calls us
at the hour when the Bride of God [the Church] rises
to sing matins to her Spouse that He may love her,
which pulls one part and thrusts another,
sounding "tin, tin" with such sweet notes
that the well-disposed spirit swells with love,
so I saw the glorious wheel move,
and voice answer voice in a harmony
and sweetness that cannot be known
except there, where joy becomes eternal.[5]

The clock image is used here to invoke feelings of religious ecstasy that are almost erotic. Various typical characteristics of the mechanical clock—the hypnotic "tin-tin" sound of the bell, the compelling rhythm of the mechanism, and the "glorious wheel" at the center—are introduced to illustrate the most exalted concept of the medieval Christian: paradise.

Especially poignant is the reference to the action of the escapement, where the one side of the escape wheel "pulls" and the other "thrusts" the two pallets of the verge, thus producing the characteristic oscillation of the foliot. The passage sounds as though Dante had been watching a clock when he wrote it.

This impression recurs when we read another passage. After visiting the Heaven of the Fixed Stars, the home of the souls of the Elect of the Church, especially apostles and evangelists, Beatrice, Dante's guide, appeals to the assembled spirits to let him advance to the highest heaven, the Empyrean, to which he will be admitted after an examination in the Christian virtues. After Beatrice's plea, the twenty-fourth canto continues:

Thus Beatrice; and those happy souls
made spheres of themselves
on fixed poles, flaming like comets,
and, as wheels in clocks turn,
so that to someone looking
one appears still, while another flies,
so those lights, dancing
at varying speeds, slow and fast,
made me judge of their riches [joy].[6]

The dances of the blessed spirits are compared to the complex but coordinated movements of a clock's gears. We can recognize in the wheel that "appears still" the wheel of the hour hand and in the other that "flies" perhaps the driving wheel of the striking train.

The two passages evaluate the clock in terms of traditional European values. When Dante, the greatest poet of the age, watched a mechanical clock in action, it made him think of the dances of the blessed spirits in paradise. Could there be a stronger endorsement for the new machine?

The clock also appeared as a literary device in the titles of some books of devotion written in the early fourteenth century in the southwest corner of Germany in the context of a movement of religious mysticism. What these theological horologia had in common with clocks was their division into twenty-four chapters, a practice not uncommon in medieval theological treatises, modeled perhaps after the *Collectiones patrum* of the church father Joannes Cassianus (360–435).[7] To the author of one of these, the *Horologium devotionis circa vitam Christi,* a Dominican monk known as Berthold of Freiburg (first half of the fourteenth century),[8] the interest in clocks extended no further. He wrote in his preface:

> I wish to call this little book *horologium devotionis* for this reason: Just like the natural day has 24 hours, counting day and night together, so this booklet of the Life of Christ has 24 chapters. . . . And I have listed the chapters of the book under the name of hours.[9]

A similar book, called *Horologium aeternae sapientiae,* was written presumably in the 1330s by Henricus Suso (or Heinrich Seuse, ca. 1295–1366), a Dominican like Berthold and a widely revered mystic.[10] Suso's use of the clock metaphor was more differentiated. The purpose of his book, he said in his preface, was not to inform the ignorant but to rekindle extinct flames of faith and to admonish the lukewarm, to provoke the impious to devotion and to stir into watchful virtue those who were lying torpid in mindless sleep.

> Hence the present little work tries to expound the Saviour's mercy as in a vision, using the metaphor of a pretty clock decked with fine wheels, and of a dulcet chime giving forth a sweet and heavenly sound, exalting the hearts of all by its complex beauty.[11]

The intended effect of Suso's *Horologium* then was, as Lynn White has put it, somewhat that of a spiritual alarm clock.

Suso did not explicitly return to the clock after the preface, but in another way its image remained present throughout the book: the body of the *Horologium sapientiae* consisted of a dialogue between the author and the personified virtue of Sapientia, or Divine Wisdom. In the popular mind, this virtue was gradually becoming associated with the mechanical clock. Illuminated manuscripts of Suso's book frequently featured the allegorical figure of Sapientia surrounded by clocks (figure 2–1); these illustrations, incidentally, show considerable technical detail and are thus of high value as sources for the history of horology.[12]

Figure 2–1 "Sapientia" amidst a selection of timepieces; illumination in a manuscript copy of Henricus Suso's *Horologium aeternae sapientiae,* circa 1450.

Berthold's *Horologium devotionis* and, even more, Suso's *Horologium sapientiae* came to be widely distributed. The demand for the first was still large enough after the invention of printing to give rise to twelve incunabula editions.[13] Suso's book, soon translated into the major European languages, is said to have been, by the end of the fifteenth century, second in popularity only to Thomas à Kempis's *Imitation of Christ;* the extant manuscripts of it alone number nearly two hundred.[14]

Later in the fourteenth century, the clock metaphor surfaced in the writings of several authors working in or around Paris. Jean Froissart (ca. 1337–ca. 1404), a Belgian priest who spent his life at the courts of various princes and kings in northwestern Europe, is best known for his chronicle of the Hundred Years' War between England and France. He also wrote a good deal of poetry, including a number of poems on chivalrous love with titles like *Le Paradis d'amour, L'Espinette amoureux,* and *La Prison amoureuse.*[15] One of his longest (1,174 lines) is *Li Orloge amoureus* (written probably in the 1360s or early 1370s), an elaborate allegory that places the various aspects of formalized courtly love in analogy with the workings of the mechanical clock.[16] Virtues and emotions like honor, loyalty, humility, and patience are portrayed through mechanical counterparts in the clock. The beauty of a lady, for example, causes in the heart of a lover desire, like the leaden weight that makes a clock run.[17]

In this manner, the analysis of the clock of love proceeds, in great detail, from the weight to the cord, from there to the "first wheel," and so forth. Noteworthy is the treatment of the central element of the clock, the verge-and-foliot escapement. It becomes the allegorical equivalent of the virtue of *mesure* (temperance, moderation, self-control). Love without *mesure* would lead to uncivilized, destructive passion, just as in a clock without escapement the force of the weight would cause an uncontrolled, violent motion:

> And, because it [the first wheel] would go without governance
> And all too swiftly and without restraint [*mesure*]
> Had it not something that from its unruliness [*desmesure*]
> Should hinder it and bring it back to rule,
> And by its due control should regulate it,
> For this, by proper art arranged,
> There was a second wheel [the escape wheel] adjusted,
> Which slows it down and makes it move
> By governance and with restraint [*mesure*], to wit
> By virtue of the foliot as well,
> The which without cessation thus is moved:
> One stroke to the right and then one to the left,
> Nor may it nor it cannot be at rest;
> Because by it this wheel is checked
> And by true moderation [*mesure*] stayed.[18]

The passage is not without merit as a technical description of the functioning of the escapement. The reference to the characteristic reciprocating motion of the verge and foliot,

> which without cessation thus is moved:
> One stroke to the right, and then one to the left,

reminiscent of Dante's "which pulls one part and thrusts another," highlights the source of a clock's motion, the crucial interaction between verge and escape wheel.

More significant than such technical detail is Froissart's comparison of the escapement with the virtue of *mesure.* Among the traditional seven virtues—the three *theological* virtues (faith, hope, and charity) and the four *cardinal* virtues (prudence, courage, temperance, and justice)—*mesure,* also known as *misura, mâze,* or *temperantia,* had in the late thirteenth century emerged as preeminent.[19] This rise probably had to do with the conventions of chivalry, which as a social and cultural movement was then reaching its most refined and formalized stage. *Mesure* or *temperantia* was the virtue that defined the perfect knight. Froissart's identification of the clock's escapement with temperance has special weight, for Froissart is regarded as chivalry's outstanding chronicler and poet. The comparison must not be overrated, however, for Froissart presented it casually and playfully, intermingled with many other metaphors and without special emphasis.

The clock-*temperantia* comparison figured more prominently in a book written in the same cultural environment about a generation later. Christine de Pisan's *L'Epître d'Othéa* belonged to a class of books that enjoyed continuing demand—treatises of manners and morals written for the benefit of adolescent aristocrats. The author (1364–after 1430), daughter of an Italian physician and astrologer at the court of King Charles V of France, was an early-widowed *femme-de-lettres* in the periphery of the French court, who made ends meet with a vigorous literary production. The *L'Epître d'Othéa,* presumably composed about 1400 for the son of Duke Philippe le Hardi of Burgundy, presented its lessons in mythological guise as a letter from the Trojan goddess Othéa to the young Hector. The clock metaphor was introduced at the start. Presenting Othéa as "the wisdom of women," Christine went on to say that

> Temperance should be called a goddess likewise. And because our human body is made up of many parts and should be regulated by reason, it may be represented as a clock in which there are several wheels and measures. And just as the clock is worth nothing unless it is regulated, so our human body does not work unless Temperance orders it.[20]

Significant here is not only the linking of the clock with the virtue of temperance but also the presence, in the background of the picture, of wisdom ("wisdom of women") and reason ("regulated by reason"). This association of temperance with wisdom had a special significance in the iconography of that day. Copies of Suso's widely read *Horologium sapientiae* were often illustrated with illuminations of Lady Sapientia shown next to mechanical clocks.[21] Early manuscript copies of *L'Epître d'Othéa* reflect this iconographic tradition, featuring illuminations of precisely the same content, with the allegorical woman's figure, now understood to be Temperantia, next to the clock (figure 2–2). Once Temperantia was acknowledged as the highest virtue, fully sanctioned by contemporary theology, Sapientia played no further role in iconography. The iconography of Temperantia, by contrast, thereafter developed independently, without further stimulation from literature. Late in the fifteenth century, artists routinely presented the seven virtues with Temperantia in the middle. The clock continued to be her chief at-

Figure 2–2 "Temperantia" adjusting a mechanical clock; illumination in a manuscript of Christine de Pisan's *L'Epître d'Othéa,* circa 1450.

tribute. In this form, as a woman with a clock, Temperantia became a familiar figure in the painting and sculpture of the fifteenth and sixteenth centuries.

At the same time, however, as Lynn White has shown, the meaning of temperance itself was transformed.[22] The shift became manifest in a new iconographic program that came into use in the middle of the fifteenth century. Now, when Temperantia stood in the midst of the other six virtues, she was marked by several additional attributes: her feet were on a windmill; strapped to her ankles were rowel spurs; and in one hand she held a pair of eyeglasses and in the other hand the reins leading to a bit and bridle placed, graphically, in her own mouth. Significantly, except for the bit, these attributes were all recent inventions. A good number of fifteenth- and sixteenth-century works of art have survived showing Temperantia with these remarkable accouterments (figure 2–3). The key to this puzzle of symbols comes from the caption under one such picture (from about 1470):

> He who is mindful of the clock
> Is punctual in all his acts.
> He who bridles his tongue
> Says naught that touches scandal.
> He who puts glasses to his eyes
> Sees better what's around him.
> Spurs show that fear
> Makes the young man mature.
> The mill which sustains our bodies
> never is immoderate.[23]

The elements of temperance enumerated here—punctuality, discretion, circumspection, caution, frugality—had little to do with the virtues of a knight and the ideals of chivalry. Instead they were, as White has pointed out, the virtues of the bourgeois townsman, describing

Figure 2–3 "Temperantia," engraving by Pieter Bruegel the Elder, circa 1560.

behavior known to be advantageous in commerce and industry. Temperance had become the basis of the ethics of the rising middle class.

While the clock symbolized, and perhaps subtly shaped, the ethics and values of late medieval society, its image had also begun to play a role in the analytical reasoning of philosophers. Nicole Oresme (ca. 1320–82) was a cleric who also held academic positions and ended as bishop of Lisieux. He was highly regarded by his contemporaries and was a personal friend and close political advisor of the shrewd and enlightened King Charles V of France. Today he ranks as one of the leading scholastic philosophers and is remembered not only for his contributions to natural philosophy and mathematics but also as an economic theorist of originality.

Oresme used clock imagery on two occasions. The earlier one was a treatise, *De commensurabilitate vel incommensurabilitate motuum celi* (probably written in the 1350s), that was designed to refute astrology.[24] In it Oresme argued that the proportions of the velocities, forces, and periods involved in the motions of various celestial bodies are incommensurable or irrational (in the mathematical sense: fractions that do not form ratios of integers) and that astrological prediction, which is based on the assumption that the orbital periods of various planets form fixed ratios, is therefore without foundation. One of the objections, he thought, that might possibly be raised against his thesis was the widespread view that incommensurable ratios were displeasing, indeed shameful, to man and nature: "For if someone should construct a material clock would he not make all the motions and wheels as nearly commensurable as possible? How much more [then] ought we to think [in this way] about that architect who, it is said, has made all things in number, weight, and measure?"[25]

Important in this objection is the premise, presumed to be shared by both sides of the debate, that the relationship between the universe and its creator is identical to that between a clock and its clockmaker. Oresme again stated this view of God's relationship to his creation in his *Livre du ciel et du monde* (1377), a commentary upon Aristotle's *De caelo.* The context here is the problem of accounting for the continuous motion of the celestial bodies. Oresme did not accept the impetus theory of his teacher Jean Buridan but adhered to the traditional belief that the celestial bodies were moved by intelligences. The evident steadiness of the motion of celestial bodies, he felt, was due to some form of equilibrium between motive powers and resistances which, in its results, reminded him of the functioning of the clock:

> These powers and resistances are different in nature and in substance from any sensible thing or quality here below. The powers against the resistances

> are moderated in such a way, so tempered, and so harmonized that the movements are made without violence; thus, violence excepted, the situation is much like that of a man making a clock and letting it run and continue its own motion by itself. In this manner did God allow the heavens to be moved continually according to the proportions of the motive powers to the resistances and according to the established order [of regularity].[26]

The notion of the Clockmaker God implied here is noteworthy with regard to both the ancient traditions upon which it rested and its subsequent influence.

To compare the world with a clock was only an extension of the ancient custom of calling the world a machine. The expression *machina mundi,* "the world machine," was used first by the Roman poet Lucretius (94–55 B.C.).[27] Subsequently, the term occurred in the works of many ancient authors (for example, Lactantius; Arnobius Afer, who opposed the notion; Firmicus Maternus; Dionysius Areopagita; Chalcidius; and Cassiodorus) and in that of most scholastic philosophers (Alanus ab Insulis, John Sacrobosco, and Robert Grosseteste).[28]

The term was generally used casually, an idiom about the meaning of which there was not expected any doubt. The expression *world machine* nevertheless carried an implication that was as far-reaching as it was inescapable, namely, that man-made machines and the divinely created cosmos had certain essential characteristics in common. For the time being, what these common characteristics were was left open—presumably the machine and the world shared an essential complexity and ingenuity of construction—but it was by no means assumed that both obeyed the same laws.

The question of creation was usually not brought up, but the assumption was implicit that the world machine, like all other machines, had been created rather than that it had existed from the beginnings of time. Some classical authors had expressed this boldly. Cicero (106–43 B.C.), who had actually seen and hugely admired Archimedes' hydraulically driven world model, repeatedly compared Archimedes' feat with that of the Creator of this world.[29] Saint Clement of Rome (ca. A.D. 100) constructed the syllogistic argument that, just as the ingenuity and beauty of a work of art is evidence of the cleverness and genius of the artist, so the order and majesty of the world prove that it was created by Divine artistry and ingenuity.[30]

This pattern of reasoning, subsequently known as the "argument from design," was to become popular one and a half millennia later as an effective method of proving the existence of God. Eventually, the argument was formulated exclusively in terms of the clock metaphor: clocks owe their existence to clockmakers; the world is a huge clock; therefore, the world, too, was made by a clockmaker—God. This line of reasoning had been anticipated by Oresme.

As the outstanding thinker of the country and a close advisor and personal friend of the king, Oresme must have been something like a center of the intellectual community of Paris. Most probably he was personally acquainted with Jean Froissart and Christine de Pisan, and these were not his only direct acquaintances for use of the clock metaphor. Henry of Langenstein (or, of Hesse, 1325–97), a German cleric and natural philosopher and an exact contemporary of Oresme, wrote, somewhat after Oresme, a scientific interpretation of Genesis in which we find formulations extremely similar to Oresme's.[31] He compared God, as Creator and First Mover, to a clockmaker who composes his machine from its many pieces and, once it is complete ("horologium totum in suis partibus, cordis et ponderibus"), sets it in motion. Elsewhere he compared the arrangement and functioning of the parts of animals to the interaction of the parts of a clock.[32] Such similarities between Langenstein and Oresme were not coincidental. From 1363 to 1382 Henry of Langenstein had worked, as a student and a teacher, at the University of Paris; Oresme, who died in 1382, always spent a major part of his time in Paris, even when assigned to various posts in Normandy. It is inconceivable, then, that Langenstein and Oresme, as clerical confreres and scholars with common interests, should not have been fairly well acquainted with each other.

With all their diversity of origin, Froissart, Oresme, Langenstein, and Christine de Pisan flourished in the same intellectual community. The details of their interactions are now impossible to reconstruct. It is safe to assume, however, that some of them, although perhaps not all four, knew one another personally and that they communicated not only through their books but also by word of mouth. Lively discussions about the mechanical clock, its wonders, and its meanings must have included some of these four (the possibility of a meeting of all four is not actually ruled out by the known facts) and also many others. In such conversations, ideas would have been shaped and modified, with different effects upon different participants, in ways for which there cannot possibly be a record. The few literary traces of the image of the clock that have survived can give us only a faint and possibly distorted glimpse of the manner and the power with which the mechanical clock fascinated the people of late medieval Europe.

In summary, this much may be said: the number of clock metaphors from the clock's first century may not have been large, but their significance is disproportionate to their number.[33] They were invoked by several of the century's leading writers—Dante, Suso, Froissart, Oresme—and their purpose was to exult society's highest values: the harmony of Paradise; chivalrous love; temperance, the chief virtue; and the divine nature of the Creation.

Flowering in the High Baroque

In the sixteenth century the metaphor began to flourish conspicuously in most of Europe. The proliferation of the metaphor was not only a result of the proliferation of the clock itself. Quite generally, the use of metaphors, symbols, mythological allusions, and other figurative devices flourished luxuriously in the literature of the sixteenth and seventeenth centuries, just as iconology, iconography, and emblematics were indispensable ancillary disciplines of the Mannerist and Baroque artists. Literary metaphors and iconographic emblems were publicized in numerous handbooks, not only for the use of practicing writers and artists but also to help the educated layman decipher symbolic riddles and generally become conversant in the fashionable sport of expressing himself on several levels of meaning at once.[34] Such handbooks of symbols, of which Filippo Picinelli's *Mundus symbolicus* was perhaps the outstanding representative, offered many entries dealing with the clock image.[35] Neither handbooks nor actual literary practice always distinguished sharply between the mechanical clock and other timekeepers such as sundials and sandglasses. The mechanical clock, for example, can be found as the iconographic attribute of Father Time (or Chronos, and sometimes Death), but more commonly this function was the prerogative of the sandglass.[36] The three types of timekeepers (clock, sundial, and sandglass) were also employed fairly indiscriminately to illustrate various general aspects of the concept, time: its fleeting nature, the importance of husbanding it, the benefits of a disciplined daily regimen, the virtue of punctuality, and so forth.[37] Since clock metaphors of this type tell us nothing about contemporary perceptions of the mechanical clock in particular, they will not be considered here.

The lowest common denominator of the clockwork metaphors in sixteenth and seventeenth century European literature was the tone of unqualified approval for the clock. Except for some seventeenth century Englishmen who were capable of viewing clocks critically, writers invoked the image of the mechanical clock consistently in support and praise of things they valued.

A more specific characteristic of most clock metaphors was the praise of order and regularity. This sentiment surfaced in short phrases like "as true as a clock," "He payes and takes as dulie as the clock strikes," "a well-framed order, clock and guide line," and in more elaborate metaphors as, for example, in *Gargantua,* where an army is described as so well trained "that they seemed rather to be a concert of organ-pipes, or a mutual concord of the wheels of a clock, than an infantry or cavalry, or army of soldiers."[38]

A characteristic feature of a clock's regularity, as the Elizabethan physician Timothy Bright (better known as the inventor of modern shorthand) observed, was that all its complex and divers motions had one single, central origin. He compared its manner of functioning to the role of the soul in the body.[39]

To illustrate a particular aspect of regularity, the English poet John Donne pointed to some rather technical aspects of the clock's behavior: prayer, he urged in one of his sermons, should be preceded by meditation and followed by "rumination," just as "a clock gives a warning before it strikes, and then there remains a sound, and a tinkling of the bell after it hath stricken."[40] His comparison was accurate: a moment before a clock strikes, it is possible to hear the striking train beginning to run.

Donne was equally technical when he tried to justify ritual and convention—normally regarded as unproductive and wasteful—as preservers of order, by comparing them to the string that connects the main spring with the fusee. Although it adds no energy, it is useful because it equalizes the driving power of the clock.[41]

John Amos Comenius, the pioneer of modern education, compared the process of well-planned systematic schooling with clockwork. The art of teaching, he explained, is no more than the mastery of time, material, and method. If this mastery is attained, then

> the whole process, too, will be as free from friction as is the movement of a clock whose motive power is supplied by the weights. It will be as pleasant to see education carried out on my plan as to look at an automatic machine of this kind, and the process will be as free from failure as are these mechanical contrivances, when skilfully made. . . . Let us therefore endeavour, in the name of the Almighty, to organise schools in such a way that in these points they may bear the greatest resemblance to a clock which is put together with the greatest skill, and is cunningly chased with the most delicate tools.[42]

Such widely shared although not sharply focused feelings of admiration for the clock were summarized by Johannes Geyger, an otherwise unknown South German Protestant pastor (1621) in his panegyric on clockwork:

> After the Supreme Governor, Director, Ruler, and Leader of all things on earth, after the heavenly lights, signs, and planets, and also after His Holy Scripture and the other beneficial regulations, statutes, and laws of the secular Christian authorities, [the clock] is the true master, moderator, ruler and guide of human life.[43]

Admiration for the clock was rooted partly in the aesthetic appeal of the regularity of its running, partly in the authority that such regularity gave it over its surroundings, and in the leadership a timekeeper exerted in human affairs. Thus, the clock became a symbol of any authority that brings order into human life. A widely read book on aristocratic morals and manners by the Spanish Jesuit Antonio de Guevara (1480?–1545?) was called *The Clock of Princes (Relox de principes,* published in English as *The Diall of Princes*). The author explained this choice of title as follows:

> This dial of princes is not of sande, nor of the sonne, nor of the houres, nor of the water, but it is the dial of life. For the other dialles serve to knowe, what houre it is in the nighte, and what houre it is of the day: but this sheweth and teacheth us, how we oughte to occupye our mindes, and how to order our lyfe. The propertye of other dyalles is, to order things publykes: but the nature of this dyal of prynces is, to teache us how to occupye our selves everye houre, and howe to amende our lyfe everye momente.[44]

The clock was compared here, somewhat as in Suso's *Horologium sapientiae,* with the personal virtues that give internal direction to the life of the responsible individual.

The comparison of clock and prince became popular as an illustration of princely authority, as a metaphor in praise and support of those who govern the life of communities. This sentiment was given straightforward expression, for example, by the early seventeenth-century London playwright, John Webster (1612):

> The lives of princes should like dyals move,
> Whose regular example is so strong,
> They make the times by them go right or wrong,[45]

by Christoph Lehmann (1630), historian, poet, and official of the city of Speyer:

> A prince and ruler is the nation's clock:
> Everyone will follow him in his conduct
> as he follows a clock in his daily affairs,[46]

and by the monarchist poet and playwright, William Davenant (1660):

> For from the Monarch's vertue *Subjects* take,
> Th' ingredients which does public vertue make.
> At this bright beam they all their Tapers light,
> And by his Diall set their motion right.[47]

From the premise that the prelates of the church and the worldly authorities are like clocks on church towers, serving to regulate the movements of others, Cesare Ripa (1618) derived the observation that the actions of persons of authority must be regulated and adjusted with great precision.[48]

The clock-prince comparison appeared in many variations. In John Donne's *Obsequies to the Lord Harrington* (1614), clock imagery was elaborated in great detail. While ordinary people were likened to pocket watches that need to be rewound and reset frequently and noblemen to church clocks that give guidance to whole communities, the subject of the eulogy was described as being of such unfailing virtue as to resemble the sundial by which the church towers themselves are adjusted.

> Yet, as in great clocks, which in steeples chime,
> Plac'd to informe whole towns, to imploy their time,
> An error doth more harme, being generall,
> When, small clocks faults, only on the wearer fall;
> So worke the faults of age, on which the eye
> Of children, servants, or the State relie.
> Why wouldst not thou then, which hadst such a soule,
> A clock so true, as might the Sunne controule,
> And daily hadst from him, who gave it thee,
> Instructions, such as it could never be
> Disordered, stay here, as a generall
> And great Sun-dyall, to have set us All?[49]

An epigram attributed by the seventeenth-century Nuremberg poet Georg Philipp Harsdörffer to Justus Lipsius, the great Flemish classicist, took the comparison a step further:

> Just as we see the hand of the clock and read the hours from its turning without having insight into the ingenious workings of its complex gears, so we can observe the blessings and punishments of God without knowing and understanding their secret causes. Likewise, the actions of princes and lords lie open before our eyes, but their purposes and motives are hidden from our eyes.[50]

Here the clock symbolized not only princely but also divine authority; the point of the metaphor was to stress the unbridgeable gulf between the sovereignty of God or king and the hopeless dependence of the subjects. Christoph Lehmann said almost the same thing: "The clock of God runs with sureness and does not fail; No ingenuity nor force will prevent this. Anyone who wishes to accomplish something, better look upon the Clock of God to find out whether the hour has struck when to perform his task."[51]

In all these examples, the nature of authority was illustrated by the picture of a tower clock with the townspeople underneath, looking up to it for guidance. The clock was presented as a higher being whose inner workings were inscrutable and mysterious. Later, as the internal mechanisms of the clock began to be more widely understood, the picture changed. The relationship between the sovereign authority and its dependent subjects was illustrated as that between the various parts of the clock, for example, between the weight and the hands: "Those who are driven by the Spirit of God and who know how to follow it, are children of God. Just like a clockwork if otherwise properly made will eagerly follow the weight."[52] Elaborate clock metaphors of this kind, where the whole of society was interpreted as a complex clockwork, soon were to play a central role in political philosophy.

Another class of clockwork metaphors that came into use in the seventeenth century—the comparison between clock and living creature—seemed to have less to do, on the surface, with regularity and order than with the popularity of automata and particularly with the obvious resemblance of the ticking of clocks with the beating of the heart: "The heartbeat is man's escapement and since a clock cannot run without one, so man's life will stop with the beating of his puls."[53] Johannes Geyger postulated the analogy between clock and body in great seriousness: "Since . . . the material clockwork of wheels, escapement, weight, springs, hands, etc., is a most miraculous contrivance: therefore it is not inappropriate to view as such the *mirabilis structura* and *fabricatio,* that is the miraculous structure of the human body."[54]

Poetic comparisons between aspects of the human body and clockwork became frequent in the seventeenth century. Their themes ranged widely, but the most popular one was the proneness of clocks to break down and the frequency of their need for repairs. An early example comes from Luis de Granada (1504–88), a Spanish Dominican and friend of Saint Theresa, who drew comfort from the observation that the clock, even though not nearly as complex, was more fragile than the human body:

> Why is a clock so oftentimes disordered and out of frame? The reason is, because it hath so manie wheeles and points, and is so full of artificial work, that although it be made of yron, yet euery little thing is able to distemper it. Nowe, how much more tender is the artificial composition of our bodies, and how much more fraile is the matter of our flesh, than is the yron whereof a clocke is made? Wherefore, if the artificial composition of our bodies be more tender, & the matter more frayle: why shoulde wee wonder if some one poynt among so manie wheeles haue some impediment, by reason of which defect, it stoppeth and endeth the course of our life? Truly we haue rather good cause to maruell, not why men do so quickly ende their liues, but how they endure so long, the workemanshippe of their bodies being so tender, & the matter and stuff whereof they bee compounded, so frayle and weake.[55]

The observation was echoed by Thomas Tymme:

> And how, I pray you, commeth it to passe, that clocks are so easily stopped from their course: is it not because they are made by arte and skill, with so many wheeles, that if one be staide, all the rest are letted? If this befall clockes that have wheeles of Iron and Staele, how much more easily may it come to passe in the humane clock of life: the wheeles and engines whereof, are not if Iron, but of clay. Therefore let us not wonder at the frailtie of mans body, but at the foolishnes of mans minde, which upon so fraile a foundation is wont to erect and builde such lofty Towers.[56]

Some authors turned the fragility of clocks into a source of reassurance; after all, clocks could be repaired. John Donne introduced this notion into a funeral elegy for a young woman (1610):

> But must we say she's dead? may't not be said
> That as a sundered clock is piecemeal laid,
> Not to be lost, but by the maker's hand
> Repolished, without error then to stand.[57]

The notion is encountered, in very similar form, on an epitaph in Exeter Cathedral (1614) and in a funeral poem composed in Nuremberg (1681).[58]

The comparison of a person's death with a clock's being out of service for repairs had numerous variations. The playwright John Webster invoked the image where "some curious artist takes in sunder a clock or watch, when it is out of frame, to bring't in better order,"[59] not in a case of death but in the parting of lovers. Bishop Henry King described the death of King Gustavus Adolphus as the unwinding of a watch (the image is complicated by an infelicitous reference to the fusee and its string):

> Thy task is done; as in a Watch the spring
> Wound to the height, relaxes with the string:
> So thy steel nerves of conquest, from their steep
> Ascent declin'd, lie slackt in thy last sleep.[60]

George Herbert compared clock repair to the process of sleep, and Sir William Pelham looked upon the winding of clocks as much the same thing as eating.[61]

The habit of Baroque writers to draw metaphorical comparisons between the clock and the living body may seem whimsical and sometimes crude. To appreciate its significance, however, we must recall that the habit was shared by contemporary natural philosophers who, by greatly elaborating the analogies between clock and body, created a new physiological theory.[62] In that theory, the clockwork qualities of order and authority were distinctly manifest.

The main themes of the clock metaphor identified so far—order, authority, and the living organism—quite naturally all converge into a fourth one. The regularity of nature, the creation of life, and the ultimate authority—all of these are concepts that imply God. God and things Divine had been linked in diverse ways with the clock image from the beginning, to recall only Dante, Suso, and Oresme. As the clock metaphor came to be used more frequently, it continued to illustrate a variety of different aspects of God. One theme began to stand out: the relationship between clock and clockmaker was taken as an accurate analogy for that between the creation and its creator. Oresme, late in the fourteenth century, had been the first to use the clock metaphor in this sense. Joachim Rheticus, in his *First Report* (1540) on Copernicus's new world system, introduced it as an argument in favor of the earth's motion around a fixed sun:

> Since we see that this one motion of the earth satisfies an almost infinite number of appearances, should we not attribute to God, the creator of nature, that skill which we observe in the common makers of clocks? For they carefully avoid inserting in the mechanism any superfluous wheel or any whose function could be served better by another with a slight change of position.[63]

This sentence was repeated a century later, with obvious approval, by John Wilkins, an Anglican bishop, son-in-law of Oliver Cromwell, and a founding member of the Royal Society.[64]

At times, the analogy was modified. The Spanish Dominican Luis de Granada compared the relationship between the world and God not to that between clock and maker but to that between the physical mechanism and its performance:

> It appeareth, that like as all the mouing and order of a clock, dependeth of the wheeles that doe drawe it and make it goe, insomuch that if they should staie, immediatlie all the whole frame and mouing of the clocke would stay also: euen so all the workmanship of this great frame of the worlde, dependeth whollie of the prouidence of Almightie God, in such sorte, that if his diuine providence should faile, all the rest would faile out of hand withall.[65]

A little later, the analogy also surfaced in France, in the writing of Philippe de Mornay, a leading Huguenot (1581):

> Sure, the sky is as the great Wheele of a Clocke, which sheweth the Planets, the Signes, the Houres, and the Tides, euery one in their time, and which seemeth to bee his chiefe wonder, proueth it to bee subiect to time, yea, and to be the very instrument of time. Now, seeing it is an instrument, there is a worker that putteth him to use, a Clock-keeper that ruleth him, a Minde that was the first producer of his mouing. For euery instrument, how moueable soever it bee, is but a dead thing, so farre forth as it is but an instrument, if it haue not life and mouing from some other thing than it selfe. . . . O man, the

> same worke-master, which hath set up the clocke of the heart for halfe a score yeeres, hath also set up this huge engine of the skies for certaine thousand of yeeres. Great are his circuits, and small are thine, and yet when thou has accounted them thoroughly, they come both to one.[66]

God was here described as the maker and keeper not only of the large clock of the universe but also of the small one of the human heart.

John Norden, a minor Elizabethan poet, extended the limits of the clockwork universe with God as prime mover even further. In addition to the physical world, he included the realm of social life (1614):

> This mouing world, may well resembled be,
> T'a Jacke, or Watch, or Clock, or to all three:
> For, as they moue, by weights, or springs, and wheeles,
> And euery mouer, others mouer feeles,
> So doe the states, of men of all degrees,
> Moue from the lowest to the highest fees,
> The lesser wheeles haue most celeritie,
> The greatest moue with farre more constancie,
> And if there mouings lowest wheeles neglect,
> The greatest mouer doth them all correct.[67]

In Germany, similarly, it had become customary to speak of the "mighty master clockmaker" of the celestial clockwork. Geyger declared in his *Horologium politicum* (1621):

> How much wiser, more understanding and inventive [than that wise, understanding and ingenious master who builds complex clocks] must be this Master who has created the master clockmaker himself, and has given him the intelligence and skill to make clockwork. Indeed he who by his omniscience has created the whole heavenly firmament and clockwork, sun, moon, planets, constellations, stars, etc., and who, to this hour and minute, by his omnipotence and wisdom maintains these in running order, he is none other than the Lord.[68]

The image of the Clockmaker God was common in such general expressions of piety. At the same time it also began to play a more serious role in theological reasoning. The clockmaker-God analogy was the basis of a formal argument that soon came to be regarded as the most compelling proof of the existence of God. But as theologians refined this argument—today referred to as the "argument from design"—they discovered that it contained a disturbing dilemma: it implied that God could not, at the same time, be both all-wise and all-powerful.

John Robinson (1575–1625), an English nonconformist divine known as "the pastor of the Pilgrim fathers," found the analogy of clockmaker and God unobjectionable as long as it concentrated on the process of creation:

> Neither doth this [delicacy of interactions between stars and earthly phenomena] at all diminish, or detract from the honour of the Lord in governing the World, but rather amplifieth it; as it addes to the honour of the skilfull Artificer, so at the first to frame his Clocke or other worke of like curious devise, as that the severall parts should constantly move, and order each other in infinite varietie, hee, as the Maker, and first Mover moving and ordering all.[69]

The difficulty arose after the work of creation was finished. If God's creation was indeed perfect, then it would function thereafter without further divine help, thus condemning God to the role of an idle spectator. To those who believed in God's omnipotence, this possibility was appalling. Robinson rejected it without debate:

> Where yet this difference must alwayes be minded, that the Artisan leaves his worke being once framed to it selfe; but God by continuall influx preserves, and orders both the *being*, and *motions* of all Creatures. Here also we except both *unnaturall* accidents; and specially, supernaturall, and miraculous events; which are, as it were, so many particular creations, by the immediate hand of God.[70]

Thus, for the time being, the troublesome question was suppressed. In the long run, however, it refused to remain unanswered. Within a few decades, the dilemma became the subject of a vigorous debate, with clock and clockmaker playing central roles as analogies of creation and creator. Ultimately, it was decided that the root of the problem lay not in the nature of God but only in the nature of the design argument or, indeed, of the clock analogy itself.

In the literatures of Continental Europe prior to the mid-eighteenth century, it is virtually impossible to find metaphors or any other comments on the mechanical clock that were in any sense negative. The rare instances of scepticism that come to mind—the episode of the clockless abbey in *Gargantua*[71] and the anecdote of Emperor Charles V illustrating the futility of any effort to establish harmony among men by the impossibility of getting two clocks to run in synchronism[72]—are not only subdued in their criticism; the latter one is also apocryphal.

The mood was quite different in England. To be sure, there were many English authors, especially the mechanistic natural philosophers around Robert Boyle, who regarded the clock as favorably as their Continental contemporaries. But, in sixteenth- and seventeenth-century England, there was also a significant body of negative comment on clocks that ranged from wry mockery and good-natured ridicule to outspoken hostility. Even the outburst of the fourteenth-century Celtic bard Dafydd ap Gwilym ("Woe to the black-faced clock . . . A curse on its head"),[73] an isolated reaction to the advent of the first tower clocks, is without counterpart in the Continental literature.

By the late sixteenth century, a pattern was beginning to emerge in English use of clock imagery. Clocks were now called not only "true" and "punctual" but also "cold," "gloomy," and "long-faced," as well as discordant and dishonest.[74] Various English authors reiterated this point: "The preachers of England begin to strike and agree like the Clocks of England, that neuer meete iumpe on a point together"; "He will lie cheaper than any beggar, and louder than most clocks"; "They agree like the clocks of London"; and "Alas, we scarce live long enough to try / Whether a true made clock run right, or lie."[75]

The most obvious defect of the clock, then, was its inaccuracy, sometimes interpreted as mendacity, a defect that advertized itself forcefully to anyone who stood within earshot of more than one clock capable of striking the hours. But the dislike of clocks had also deeper roots.

The various automata figures with their clumsy, ineffectual motions and their lack of a will of their own were convenient models for disparaging comparisons. The simplest of these, "jacks," were male figures with hammers that struck the bells of large public clocks. Unkind comments about clock-jacks and automata figures were common in seventeenth-century English literature: "What is mirth in me, is as harmless as the quarter-jacks in Paul's, that are up with their elbows four times an hour, and yet misuse no creature living"; "This is the night, nine the hour, and I the jack that gives warning"; "I think [said a defeated soldier describing the enemy] by some odd gimmors or device Their arms are set, like clocks, still to strike on; Else n'er could they hold out so as they do"; "He looks like a colonel of Pigmies horse, or one of these motions in a great antique clock"; and "that old Emerit [hermit] thing, that like / An Image in a German clock, doth move, / Not walke, I mean that rotten Antiquary."[76]

Englishmen also found it amusing to use the clock as an unflattering illustration of certain allegedly feminine characteristics.[77] Of special significance is an example provided in Shakespeare's *Love's Labors Lost* (1594–97) where someone, on the suggestion that he might be in love, bursts out:

> What! I love, I sue, I seek a wife—
> A woman, that is like a German clock,
> Still a-repairing, ever out of frame,
> And never going aright, being a watch,
> But being watch'd that it may still go right![78]

The theme was repeated by a number of subsequent playwrights:

> No German Clock nor Mathematicall/
> Ingin whatsoeuer requires so much reparation
> as a womans face. . . .[79]
>
> What is she [woman], took asunder from her clothes?
> Being ready [dressed], she consists of an hundred pieces,
> Much like your German clock, and near ally'd;
> Both are so nice, they cannot go for pride:
> Besides a greater fault, but too well known,
> They strike to ten, when they should stop at one.[80]
>
> She takes herself asunder still when she goest to bed,
> into some twenty boxes; and about next day noon is put
> together again, like a great German clock: and so comes
> forth, and rings a tedious larum to the whole house,
> but for her quarters.[81]

The frequency with which the disliked clocks were identified as German is significant. It is true, of course, that Germany at the time was Europe's chief producer of clocks and that the flowering of clockmaking in England was still one or two generations away.[82] The passion with which Englishmen rejected whatever the mechanical clock stood for, however, is remarkable. Anything as uncongenial as that had to be foreign.

English authors who reflected more deeply understood well enough that their dislike of the clock was in part the familiar resentment of bearers of ill news: what made clocks exasperating was their evident tendency to run fastest when time was most precious and their tactless insistence on announcing the hours when that service was not welcome.[83] This observation was expressed in many styles and moods, ranging from light-hearted witticism to high tragedy. According to Robert Greene, for instance, "The Usurers Clocke is the swiftest Clock in all the Towne; tis, sir, like a womans tongue, it goes euer halfe an houre before the time; for when we were gone from him, other Clocks in the Towne strooke foure."[84] Other authors echo similar sentiments: "Laboring men count the clock oftenest . . . are glad when their task's ended," and "The clock upbraids me with the waste of time."[85] And Michael Drayton has said:

> I hate him who the first Devisor was
> Of this same foolish thing, the Hower-glasse,
> And of the Watch, whose dribbling sands and Wheele,
> With their slow stroakes, make me too much to feele
> Your slackenesse hither, O how I doe ban,

Him that these Dialls against walles began,
Whose Snayly motion of the mooving hand,
(Although it goe) yet seeme to me to stand;
As though at *Adam* it had first set out,
And had been stealing all this time about,
And when it backe to the first point should come,
It shall be then just at the generall Doome.[86]

The outburst of the impatient lover against the clock is self-mocking, but Prince Hal's lecture on Falstaff's conduct, although superficially lightened by comical clockwork imagery, is deadly serious in its unease about the phenomenon of time:

FALSTAFF: Now, Hal, what time of day is it, lad?
PRINCE: Thou art so fat-witted with drinking of old sack, and unbuttoning thee after supper, and sleeping upon benches after noon, that thou hast forgotten to demand that truly which thou wouldest truly know. What a devil hast thou to do with the time of day? Unless hours were cups of sack, and minutes capons, and clocks the tongues of bawds, and dials the sign of leaping-houses, and the blessed Sun himself a fair hot wench in flame-color'd taffeta; I see no reason why thou shouldst be so superfluous to demand the time of day.[87]

Different in mood and style but equivalent in their basic tenor are these simple lines by John Davies of Hereford:

Whereas I heare Times sober Tongue (the clock)
Call on me eu'rie howre to minde mine end,
It strikes my hart with feare at eu'rie stroke
Because so ill Time, Life, and Breath, I spend.[88]

Applying elaborate clockwork imagery to the same theme, Shakespeare achieved a dramatic climax in the last monologue of *Richard II,* where the protagonist reflects upon a failed life:

I wasted time, and now doth time waste me;
For now hath time made me his numb'ring clock:
My thoughts are minutes, and with sighs they jar
Their watches on into mine eyes, the outward watch,
Whereto my finger, like a dial's point,
Is pointing still, in cleansing them from tears.
Now sir, the sound that tells what hour it is
Are clamorous groans, that strike upon my heart
Which is the bell. So sighs, and tears, and groans
Show minutes, times, and hours; but my time
Runs posting on in Bullingbrook's proud joy,
While I stand fooling here, his jack of the clock.[89]

The reasons for the clock's unpopularity are thus gradually exposed: by mercilessly counting out to us the elapsing of our lifetime, it reminds us of the objective limits that are placed on our subjective desires and of the larger, irresistible forces that curb our freedom.

The clock delivered this message to Continental Europeans and to Englishmen alike, but there was a difference in how the message was received. Continental authors, in general, took it as a wholesome moral and as an incentive for purposeful activity; for many Englishmen, by contrast, it was a reminder of one of life's more somber truths, of a reality that one had to accept but that one could not be forced to like.

Not all British witnesses, of course, viewed the clock as a symbol with negative connotations. Quotations from others who equated the clock image with their highest values have been presented in this chapter, and more will follow. It is difficult to estimate quantitatively which evaluation of the image, positive or negative, was more widely spread in Britain. On the Continent, however, there is no evidence of such a split; there the clock symbolized only whatever was good. The English custom of deprecating the clock image thus becomes significant. It may be registered as an early indicator of an intellectual movement that later was to become powerful (see chap. 6).

In the general literature of Europe, the clock metaphor reached a peak of popularity and freshness sometime around the middle of the seventeenth century. Several themes had emerged to which the metaphor was principally devoted: the praise of order and regularity, the exultation of authority, the construction of the animal body, and the character of the created world. These themes were subsequently discussed with particular seriousness, always with the clock metaphor in the center, in two special literatures, those of natural and of political philosophy. These discussions will be examined more closely in the following chapters. In the general literature, the clock metaphor remained as popular as ever but more and more was cast in the role of a conventional rhetorical ornament, devoid of freshness and graphic force.

3

The Clockwork Universe

The Mechanical Philosophy

To say that the clock metaphor, after the High Baroque period, remained significant only within the confines of some special literatures, while not incorrect, would hardly do justice to historic reality. The debate about the nature of the world and man's relationship to it, which occupied Copernicus, Bacon, Galileo, Kepler, Descartes, Boyle, Leibniz, Newton, and countless others, amounted to more than a special literature: It was an extraordinary intellectual movement—now known as the Scientific Revolution—that not only reshaped man's view of nature but also interacted powerfully with every major factor of European civilization, be it theology, technology, or politics. A central characteristic of the Scientific Revolution was the commitment of its participants to thinking "mechanically," and although the word *mechanical* had many shades of meaning, whenever it was to be illustrated by a concrete mechanism, the choice was usually a clock. In the conceptual make-up of the Scientific Revolution, therefore, the clock metaphor was an important ingredient.

It is this feature that makes the Scientific Revolution interesting from the point of view of the present study. This chapter will proceed like the last one, that is, it will analyze clock metaphors for their implied meanings in order to gain insight into the connections between public mentalities and the clock's mechanism and technology.

Before this task is begun, however, a few words are in order on mechanical aspects of the Scientific Revolution and on the diversity of meaning of the word *mechanical.* Scientific Revolution is a modern expression. Contemporaries of that era did not have an agreed-upon name for it. They described it variously as the "experimental," "atomic," "corpuscularian," "mechanical," or simply the "new" science (or "philosophy," or "hypothesis").[1] In the view of Robert Boyle, the "great father figure of British natural philosophy of his time,"[2] it was "the philosophy, which is most in request among the modern virtuosi, and which by some is called the new, by others the corpuscularian, by others the real, by others (though not properly) the atomical, and by others again the Cartesian (or mechanical) philosophy."[3] And as the champions of the new philosophy differed among each other on many points, they shared in no consensus on what these adjectives meant; some treated them as synonymous, others as labels for rival factions.

But, to indicate the spirit of the overall movement, the word used perhaps most frequently was *mechanical.* "All recent philosophers want to explain the physical world in the mechanical manner," observed Gottfried Wilhelm Leibniz.[4] Even those who refused to be labeled as *mechanical philosophers* used the word freely to depict certain characteristics of the new science.

The character of the new *mechanical philosophy* was described in a wide range of ways. Above all, this term signaled a deliberate break with the traditional scholastic philosophy, with metaphysical speculation, and with the ancient custom of interpreting nature through magic. "After I had freed myself from the ordinary school philosophy," recounted Leibniz, "I fell upon the writings of the Moderns . . . at age 15 . . . I pondered whether I should retain the substantial forms. Mechanism finally won out and led me to devote myself to mathematics."[5] Francis Bacon before him had characterized the contrast even more pointedly:

> The operative doctrine concerning nature I will likewise divide into two parts, and that by a kind of necessity, for this division is subject to the former division of the speculative doctrine; and as Physic and the inquisition of Efficient and Material causes produced Mechanic, so Metaphysic and the inquisition of Forms produces Magic.[6]

Mechanical, then, in a certain sense was the opposite of *magical.*[7]

Like many revolutionary movements, the mechanical philosophy bore a label that at first was derogatory. In social estimation, the mechanical arts generally ranked below the liberal arts; the liberal arts connoted gentility, the mechanical arts vulgarity.[8] In Shakespearean English, "mechanick" was an insult (for further examples, see chap. 6).[9] By adopting the despised label "mechanical," the new philosophers were partly demonstrating their break with sterile traditions and partly signaling a democratic solidarity with the lowly class of craftsmen and artisans. "The most acute and ingenious part of men being by custom and education in empty speculations, the improvement of useful arts was left to the meaner sort of people, who had weaker parts and less opportunity to do it, and were therefore branded with the disgraceful name of mechanics," wrote Thomas Sydenham, an English physician and Baconian.[10]

While rejecting the "essences," "substantial forms," and occult forces of scholastic philosophy, the advocates of this mechanical philosophy stressed their commitment to explaining nature in terms of quantities that could be counted and measured. They were fond of quoting the phrase from The Wisdom of Solomon (11:20) that God "has disposed all things by measure and number and weight," perhaps in the not uncommon desire of revolutionaries to justify their radical designs in terms of traditional values.[11] The triad of "measure, number, and weight" continued to be used in definitions of mechanical philosophy but often was expanded with references to motion (Galileo, Hobbes, and Keill).[12] Later such formulas were reduced to the brief rule, repeatedly offered, for example, by Boyle: "there cannot be fewer principles than the two grand ones of mechanical philosophy, matter and motion."[13] But this rule was not very descriptive with regard to the nature of motion. Thus, it was supplemented by another definition that was simpler and more tangible and that was to remain in use for centuries: the hallmark of the new philosophy was the claim that it analyzed the phenomena of nature as though they were actions of machinery. Early suggestions of this method can be found in the writings of Francis Bacon;[14] in the philosophy of René Descartes, it played a central role (see the next section of this chapter).

Robert Boyle, in this regard a Cartesian who was exceptionally interested in matters of definition, endorsed the machine analogy on many occasions. To any "person of a reconciling disposition," he observed, the different factions of the new science, be they called atomical, Cartesian, corpuscular, particularian, or Phoenician, all alike consisted in giving "an account of the phaenomena of nature by the motion and other affectations of the minute particles of matter. Which because they are obvious and very powerful in mechanical engines, I sometimes also term it the mechanical hypothesis or philosophy."[15] The central feature of this mechanical hypothesis was its predilection for interpreting the world through the model of the machine. To Boyle, it was a

> hypothesis, supposing the whole universe (the soul of man excepted) to be but a great Automaton, or self-moving engine, wherein all things are performed by the bare motion (or rest), the size, the shape, and the situation, or texture of the parts of the universal matter it consists of; . . . So that the world being but, as it were, a great piece of clockwork, the naturalist, as such, is but a mechanician; however the parts of the engine, he considers, be some of them much larger, and some of them much minuter, than those of clock or watches.[16]

This definition endured. Johann Christoph Gottsched, a German philosopher and poet, believed, much like his teacher Christian Wolff, that "whoever looks upon the world as such a machine . . . practices mechanical philosophy."[17] Immanuel Kant stated similarly, "The method of explaining the specific differences of materials through the constitution and composition of their smallest particles in terms of machines is the mechanical philosophy of nature."[18]

For the mechanical philosophers, then, the world was a machine. To some of them, perhaps, machine was no more than an abstract concept. Those who preferred a concrete model could and did choose from the wide range of machines employed by the technology of that time.[19] But when one surveys the mechanical models, metaphors, and images actually used, one quickly discovers that those based on clocks outnumber all others.

One other aspect of the mechanical philosophy is important. Many of its spokesmen were Christian theologians, Catholics as well as members of the various Protestant denominations, and most of the scientists who advocated this philosophy were not only devout Christians but also active participants in theological debates. To a large extent, then, the mechanical philosophy was a theological issue.[20] The mechanists' interest in theology may have contained an element of self-defense: there were, in the eyes of many, only small steps from "mechanism" to "materialism" to "atheism," and critics were inclined to equate all three. It was to disarm such charges, perhaps, that the mechanists often took the offensive by arguing that their philosophy could lend valuable support to Christian orthodoxy. Specifically, they offered to mediate between traditional teachings and the various modern intellectual movements, such as the new science, on one hand, and a wide range of theological innovations, on the other. The chief contribution that the mechanists had to offer to the discussion was the "argument from design," which, as mentioned earlier, claimed to prove the existence of God by comparing the relationship between God and the created world with that between a clockmaker and a clock. This argument, either fully articulated or in various looser formulations, runs through the writings of mechanical philosophy like a continuous thread.[21]

The diverse connotations of the word *mechanical* in seventeenth-century natural philosophy suggest in rough outlines the relationship between the clock metaphor and the Scientific Revolution. The relationship can perhaps be summarized as follows: within the complex and heterogeneous movement known by that name, the mechanistic philosophy was a central element. Of this mechanistic philosophy, in turn, it was a central principle to picture the phenomena of nature as similar or equivalent to actions of machines. To mechanical philosophers, finally, as to most of their contemporaries, the archetypical machine was the clock.

What was the root of this fascination with clocks? The Scientific Revolution is not a concern of this study as such, but it promises to be a rich source of material that might help us understand the extraordinary appeal of mechanical clocks to early modern Europeans. Accordingly, this chapter proposes to present a survey of clock metaphors from the literature of the Scientific Revolution, ordered pragmatically in such categories as offer themselves, and otherwise chronologically.

The clock metaphor, unlike the mechanical mode of explicating natural phenomena in general, became popular in natural philosophy only at a surprisingly late stage. Attempts to explain the world mechanically, however, can be traced, at least as a minor thread among the developments, back to classical antiquity. The attempts of various Greek astronomers to explain mechanically the motions of the planets—from the system of homocentric spheres of Eudoxos and Aristotle to the classical systems of epicycles and deferents of Apollonius, Hipparchus, and Ptolemy—were designed in sufficient geometric and kinematic detail that a good mechanic could have realized three-dimensional functioning models from them. Such models were indeed built. Several reports have survived of ancient attempts to depict the working of earth and heaven by mechanical means. Some of them were simple and naïve, like Hero's floating earth globe and Vitruvius's representation of planetary motion by means of ants running in the concentric grooves of a clay plate turned on a potter's wheel.[22] Others were more refined. Most famous was the planetarium of Archimedes known through the testimony of several classical authors. Cicero had seen it himself; according to his detailed description—only a fragment has survived—it was a machine made of bronze which, propelled by a central drive mechanism, demonstrated the sun, the moon, and five planets all revolving correctly at their proper speeds.[23] Elsewhere Cicero told of a similar planetarium built by Posidonius of Apamea, and some three centuries later the church father Eusebius reported a planetarium clock in Palestine.[24] Other ancient mechanical representations of celestial events were the computing device of Antikythera and the planispheres of Hipparchus and Ptolemy,[25] forerunners of the astrolabe, which could be linked with water clocks to produce the earliest astronomical timepieces.

As already recounted in chapter 1, the tradition of these ancient astronomical mechanisms, transmitted through the planetariums, astronomical water clocks, and astrolabes of the Islamic and Christian Middle Ages, contributed directly to the technology of the mechanical clock. Almost all early mechanical clocks had a minimum of astronomical capabilities, but specialized astronomical clocks were capable of reproducing almost any celestial phenomenon. Such astronomical clocks were built, starting with the mid-fourteenth-century machines of Richard of Wallingford, Giovanni de' Dondi, and that of Strasbourg Cathedral, in ever-increasing frequency. Late-sixteenth-century clockwork-driven celestial globes, notably those by Eberhard Baldewein and Jost Bürgi, were capable of replicating the heavenly phenomena from the Ptolemaic viewpoint in great precision.[26] After the turn of the seventeenth century, equivalent representations based on the Copernican system became increasingly popular.[27]

As enduring as the ancient tradition of practical astronomical mechanisms was the underlying scientific attitude. Renaissance astronomers like Peurbach, Regiomontanus, and Copernicus used the same kinematic techniques as Hipparchus and Ptolemy. What was revolutionary about Copernicus's cosmology was only its heliocentric conclusion. In style and conceptual content, it was comparable to Ptolemy's. His system and that of Copernicus have been shown to be kinematically equivalent, that is, the same planetary motions relative to the sun could be produced by either system. Both were equally well suited to mechanical representation.

While the mathematical-mechanical description of the universe and its three-dimensional depiction by clockwork-driven models advanced to unprecedented refinement, it was more and more clearly recognized that the mechanical construction of such models had nothing to do with the physical realities of the world. Speculation about natural phenomena became self-critical and cautious, favoring empirical evidence over the free roaming of the imagination. Although the slogan, the machine of the world, was commonly used, the traditional belief that the physical universe was constructed quite literally of crystalline spheres, held together and powered by wheels, gears, levers, and the like, was no longer tenable (figure 3–1). It had been seriously damaged, for example, by Tycho Brahe's observations of a comet (1577) whose travel would have had to cross planetary spheres and to break them if they had been made of crystal.[28] Various cosmological schemes were proposed that tried to match the observed phenomena in a more realistic manner. William Gilbert, whose cosmology was based upon the magnet, ridiculed the old views, asking "Who will make a wheel a mile in diameter . . . to rotate a ball the size of a fist?"[29]

Bacon likewise disparaged the use of mechanical models in astronomy as reinforcing the harmful power of a priori theories; thus, they were dangerous to the empirical attitudes upon which he insisted. He complained, for example, that in astronomy,

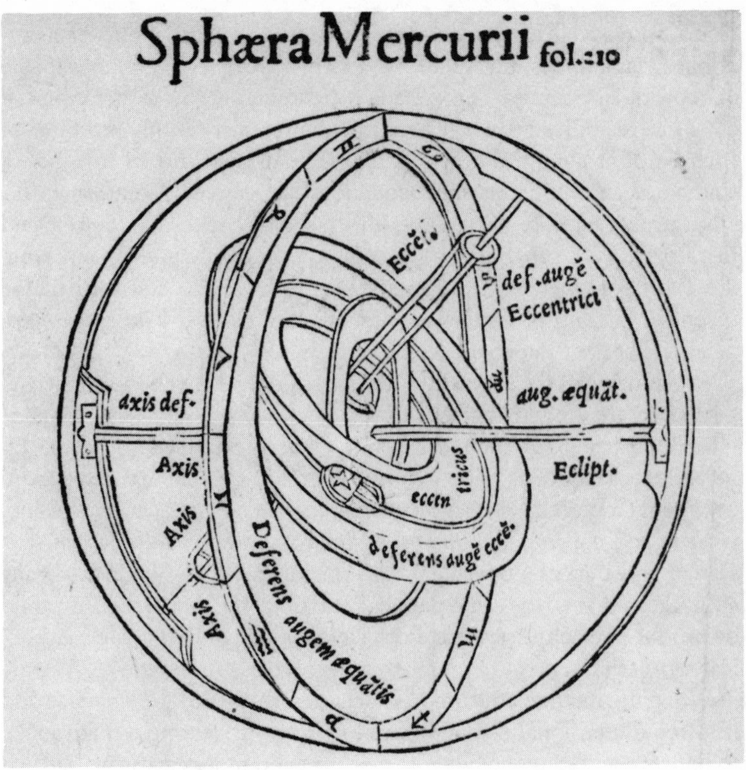

Figure 3–1. The sphere of the planet Mercury, from a sixteenth-century commentary on Georg Peurbach's *Theoricae novae planetarum* (1454).

> all the labor is spent in mathematical observations and demonstrations. Such demonstrations however only show how all these things may be ingeniously made out and disentangled, not how they may truly subsist in nature; and indicate the apparent motions only, and a system of machinery arbitrarily devised and arranged to produce them,—not the very causes and truth of things.[30]

Any surviving belief in a machinelike universe was laid to rest forever when Galileo's telescopic observations, published in the *Sidereus nuncius* (1610), established that the moon, the planets, and presumably the stars as well were objects of the same constitution and character as the earth.

By the turn of the seventeenth century, when the Scientific Revolution and with it the mechanical philosophy was about to burst into full bloom, the clock metaphor, it seems, was distinctly less popular among natural philosophers than among other writers. Famous is Johannes Kepler's programmatic statement of 1605:

> It is my goal to show that the celestial machine is not some kind of divine being but rather like a clock . . . . In this machine nearly all the various movements are caused by a single, very simple magnetic force, just as in a clock all movements are caused by a simple weight. Indeed I also show how this physical conception can be presented mathematically and geometrically.[31]

This passage, however, should not be overrated. It expresses a view he held only intermittently; at other times he preferred the traditional organic point of view.[32]

Even from a formally literary point of view, the clock metaphor was not in harmony with certain aspirations of the new science.[33] Generally, in the quest for truth about the physical nature, preference was given to empirical evidence; explanations were to be based on objective, verifiable observation. Speculation, deductions from a priori hypotheses, and reasoning by analogy were discredited.

This new outlook on the practice of science was reflected in the writing style cultivated by the new philosophers. As they committed themselves to rational reasoning on the basis of painstaking empirical observation, they cultivated a language that was artless, simple, and direct, as is exemplified in Galileo's prose. This new style was propagated with deliberate effort. Thomas Sprat has described in detail the labors, in obedience to Francis Bacon's principles, of the first members of the Royal Society of London to encourage a "close, naked, natural way of speaking" suitable for the needs of the new science.[34] They declared that they could live without metaphors ("poterimus vivere sine illis") and denounced the "vicious abundance of *Phrase,* this trick of *Metaphors,* this volubility of *Tongue*" peculiar to the language of "Wits, or Scholars" in favor of "Mathematical plainness."[35] Matching denunciations of metaphor could be compiled from many other eminent English writers of the period.[36] It can be shown, however, that these reformers were not consistent in their aims and that even their own writings showed discrepancies between precept and practice; they all used metaphors themselves when it was opportune, and clock metaphors among them.

By the mid 1600s things changed drastically. The metaphor began flourishing conspicuously in mechanical philosophy and was applied with such frequency and prominence that it must have rated as one of the shaping and driving forces of the movement. The campaigns for factuality in reasoning and sobriety in expression that had been fought against the use of models and metaphors in the new science clearly had not been successful; the image of the clock was too broadly appealing to be so easily suppressed. Nevertheless, there was one distinct historical event that helped decisively to launch the clock metaphor on its successful career in mechanical philosophy: the appearance of the philosophy of René Descartes.

Descartes

René Descartes was the quintessential mechanical philosopher. Mechanics for him was the key to the secrets of nature; to him, mechanical meant machinelike.[37] And it was the basis of his program to describe "this earth and all the visible world in general as if it were only a machine in which there is nothing to consider but the shapes and movements of its parts."[38] Past philosophies had failed, he judged, because they had not made use of the mechanical approach.[39]

The conventional rule that mechanical understanding of a thing was knowledge of the "forces, positions and shapes" involved in it (one of the many circulating variations of the Solomonic triad of "number, weight, and size") was amended now to the effect that such knowledge was more easily achieved by analyzing the thing as though it were machinery.[40] For the study of nature, Descartes postulated a far-reaching analogy between natural and man-made objects: "There are certainly no rules in Mechanics that do not also belong to Physics, of which it is a part or special case: it is no less natural for a *clock* composed of wheels to tell the time than for a tree grown out of a given seed to produce the corresponding fruit."[41]

The cause-and-effect relationships between the wheels, springs, and weights of machinery were openly visible, unambiguous, and expressible in mathematical language. It was Descartes's central belief that causal relationships of the same kind were the basis of the processes of nature.

And if, in exploring an unknown natural phenomenon, it was not possible to analyze it like a machine by taking it apart, then Descartes recommended an alternative: to invent a hypothetical process that would reproduce the same phenomenon. This would yield, of course, only probable, not certain knowledge, just as two clocks, although looking alike and performing the same function, may contain totally different internal mechanisms.[42]

The imagery in Descartes's quotations, selected only to illustrate his commitment to mechanical philosophy, shows that, if mechanical meant machinelike, machines tended to be envisioned as clocks and automata. This choice of imagery was determined not only by his theoretical philosophical objectives but also by a very personal fascination with practical technology.

Descartes's interest in mechanical technology was broad and lively.[43] His writings—published books as well as informal correspondence—

abound with references to water-lifting machines and pumps, water wheels, flying apparatus, and perpetual motion devices.[44] He had an active part in the development of a machine for grinding optical lenses and in a minor improvement of the clock.[45] But automata—"self-moving machines" under which clocks were subsumed—had a special fascination for him.[46] Descartes was personally familiar with their workings. Watches and clocks by then were economically attainable for a good part of the population, and he probably owned a watch or clock himself. Ubiquitous public tower clocks struck the hours and were often capable, in addition, of performing mechanical music, making astronomical predictions, and entertaining with automatic theater. Automata were much written and talked about. The legends of the dove of Archytas, the cosmic machine of Archimedes, the speaking head of Albertus Magnus, and the eagle of Regiomontanus were familiar to all educated people.[47] The cathedral clocks of Lübeck and Strasbourg and the mechanical miracles of the gardens of Tivoli, Pratolino, Fontainebleau, and Saint Germain-en-Laye were mandatory stations on the itinerary of the European traveler and were attentively described in many travel journals.[48]

Descartes had shared this enthusiasm for automata from his youth. In a notebook (*Cogitationes privatae*) written at age twenty-two, he described an automaton of his own invention, a tightrope walker powered by magnets, and speculated about how the dove of Archytas might be reproduced ("it will have mills between the wings, rotating in the wind"). Father N. I. Poisson, a contemporary commentator on the *Discours de la Méthode,* reported that he had seen among Descartes's manuscripts descriptions of several automata, among them not only the two mentioned above but also the design of an automaton representing a partridge being flushed out of a bush by a spaniel, "the most ingenious of them all." Poisson added, perhaps overconfidently, that Descartes's plans appeared to him sufficiently specific to be carried out without much difficulty.[49]

References to automata are frequent and prominent in Descartes's writing. There are countless casual allusions to automata that may not signify more than the author's fondness for such devices, but automata also served in deliberate analogies that contributed significantly to the substance and structure of his thought.

If Descartes, as he declared repeatedly, could "see no difference between the machines built by artisans and the various bodies composed by nature alone,"[50] what were those "bodies" in particular that he investigated with the help of the machine analogy? They were not the celestial bodies. Considering the popularity of the slogan, the world-machine, at his time and before, one might have expected that the solar planetary system would have given Descartes an opportunity to invoke clockwork analogies. In his theory of planets and stars, however, references to automaton and clockwork are conspicuously absent. Descartes's cosmology was based on a hypothesis of vortices that, while mechanical in a broader sense, did not invite or even permit direct comparison with machinery.

The "natural bodies," instead, to which Descartes applied the machine analogy so rigorously and consistently throughout all of his writings, were the bodies of animals and humans. Casual expressions like "the machine of our body" are innumerable, and his physiological writings are replete with technological analogies and comparisons to horological, hydraulic, and thermal machinery. All these comparisons were variations on a single theme: Descartes believed that animals were automata. He expressed this in many forms. He said it directly—"It is less likely that all worms, flies, caterpillars, and other animals are gifted with an immortal soul than that they move in the manner of machinery"[51]—and indirectly—"If any such machines had the organs and shape of a monkey, or of some other animal that lacked reason, we would have no way of recognizing that they were not of the same nature as these animals."[52] He repeated this argument, that it is impossible to prove animals *not* to be automata, a number of times, and humans were included in the claim.[53]

Descartes had constructed a rather novel theory of physiology of the human and animal body which he presented in numerous incidental writings and also in several longer monographic works, notably in a book written in the early 1630s but published, under the title *Traité de l'Homme,* only posthumously in 1662.[54]

Explicit comparisons with automata were introduced sparingly and cautiously.[55] Instead, in the *Traité de l'Homme,* Descartes offered a direct account of the workings of the human body which was so rigorously mechanical in all details that, in overall effect, it was equivalent to a description of a complex automaton.

Descartes took great care to avoid being interpreted as postulating any identity of human beings with automata. He used several precautions. One was the formula, exemplified in various quotations,[56] that mechanical expressions of animal or human life were equivalent to actions of hypothetical automata, but that judging by their mechanical capabilities, an actual identity between automata and living beings could not be proved or disproved. In his full-length *Traité de l'Homme,* this formula took the form, if you build a mechanism such as is described here, you will obtain a machine with capabilities that cannot be distinguished from those of the human body.

Descartes's mechanistic physiology had a second precaution: he insisted that there was one fundamental difference between man, on the one hand, and automata and animals, on the other: man's endowment with a soul. This led to the dualistic doctrine of man's two aspects: the physical aspect of the machinelike body functions and the spiritual aspect of the functions of the soul, reason, emotions, and free will. Committed to accounting for all physical phenomena in terms that were uncompromisingly mechanical, Descartes constructed an elaborate theory of the nature of the soul and of the mechanisms of its interactions with the body: the soul, as the highest in the hierarchy of organs of the human body, was located in the "pineal gland," a somewhat mysterious small organ in the brain of humans and other craniate vertebrates.[57] Floating freely in a cavity filled with "animal spirits," somewhat like "a body attached only by threads and sustained in the air by the force of fumes leaving a furnace," the pineal gland, by its changing position, regulated the supply of all bodily organs and thus of their actions with animal spirits.[58] How the soul controlled the position of the gland, however, Descartes never explained. It was this doctrine that he chiefly relied upon to draw the distinction between automaton and man and thus to prevent his mechanistic physiology from becoming classifiable as heresy; either to pacify his own scruples or to disarm the suspicions of the Church authorities, he invoked it with great frequency.

Even with all these precautions built into his theory of physiology, it seems that Descartes did not feel totally safe. Upon hearing of Galileo's difficulties with the Inquisition following the appearance of the *Dialogue concerning the Two Chief World Systems* in 1632, he withheld publication of his most extensive treatment of physiology. Written in the early 1630s as part of a larger work to be called *Le Monde,* it was published posthumously in 1662 under the title *Traité de l'Homme.*

Descartes's interest in physiology had a special orientation; content to leave to others all questions of medical and biological detail, he concentrated on the cause-and-effect relationships between the various parts of the living system. His goal was to explain the phenomenon of life through characteristic interactions between organic functioning, sense perception, and the spiritual and intellectual faculties. One might even describe Descartes's preoccupation with the total system as "cybernetic" if one recalls that it was the "control and communication in the animal and the machine" which Norbert Wiener dubbed "cybernetics" in 1948.[59]

Evidently, Descartes regarded animal organisms and automata as systems of the same kind. What were their common characteristics? To begin with, the various organs of the living body, as well as the various mechanical elements of the automaton, were ordered hierarchically on several levels in a pyramid-shaped command structure that had a single supreme organ at its vertex and a great number of subordinate organs at its base. The organs of any given level controlled the usually more numerous organs of the next level below. The organ at the vertex had sovereign power over the whole system. In humans this supreme organ was the soul with its unique capability of thought, emotion, and autonomous will. In animals it was the brain that, consisting entirely of memory and therefore capable only of initiating preprogrammed actions, corresponded closely to the mechanical programs controlling automata. Between the different levels of the hierarchy, organs were connected in linear, unidirectional, cause-and-effect, or command chains that were envisioned in entirely mechanical terms, "just as, pulling one end of a cord, one simultaneously rings a bell which hangs at the opposite end."[60]

Descartes knew that automata were capable only of actions that were engineered into them in advance or, in modern terminology, that were "program-controlled."[61] But he did not have a simple equivalent for our term *program,* just as the program function of the clocks and automata of his time was not located in a single, easily identifiable element but was distributed throughout the mechanism in the form of count wheels for the striking of the hours, pin drums for recorded music, contoured cans for automata movements, and so forth. Instead, he characterized preprogrammed automatic action as acting "according to the disposition of their organs" or "naturally and by such spring forces as a clock."[62]

The automaton's functioning according to a built-in, predetermined program accounted for its impressive capabilities as well as for its serious shortcomings.[63] With a suitable program, automata could perform astonishing feats, surpassing humans by far, but without one they could do nothing. Their capabilities were narrow and specialized; in particular, they could not make appropriate responses to unexpected challenges from the outside.[64] Automata, then, were not capable of participating in dialogue or answering questions. "Unless accidentally, automata never respond, by word and even by sign, to any questions that are put to them."[65] They were unthinking slaves of their mechanical programs and "do not have a free will." Animals, which Descartes viewed as equivalent to automata, were fully included in his ambivalent assessment of automata.[66]

Descartes assumed that lack of freedom was inherent in all mechanical devices. Convinced that animals had no free will, he saw no obstacle to a fully mechanistic physiology of animal organisms on the basis of equating the animal body with the automaton. For human beings, however, Descartes regarded the free will as indispensable, on the basis of both practical observation and ethical conviction. To account for that combination of freedom of will and machinelike body that seemed so characteristically human, Descartes depended upon his theory of the soul.

Continental Mechanists

The fortunes of the clock metaphor in the century after Descartes are best surveyed separately for the Continent (which in the seventeenth century meant the sphere of French culture) and England. Not only were these, notwithstanding a good deal of cross-channel communication, two rather independent developments; more importantly, they led to notably different outcomes.

Descartes was widely read and hotly debated. Although his influence on his own generation and on those that followed was extraordinarily broad and deep, most of his readers seemed eager to make the most of their disagreements and criticisms with him. His immediate disciples, figures of lesser repute today, while unquestionably dedicated to his teachings, entertained internal controversies of their own. The major figures, at the same time, took care to keep a safe distance between themselves and the Cartesian doctrines, even when there was significant agreement.

Three prominent younger contemporaries, Pascal, Huygens, and Spinoza, are representative. Blaise Pascal's (1629–95) incompleted *Pensées,* written at the end of his life and published posthumously, contains several strong condemnations of Descartes. One of these thoughts states baldly, "Descartes inutile et incertain," and another begins with "I cannot forgive Descartes"[67] and goes on to accuse him of virtual atheism. Pascal rejected Descartes's mechanistic interpretation of nature:

> Descartes: We must say by and large: "It is done by figure and movement," for that is a fact. But to say what, and make up the machine, is ridiculous; for that is useless and uncertain and tiresome. And even if it were true, the whole of natural philosophy is not, to my mind, worth an hour's labour. (No. 285a)

But at the same time, he used the metaphors of the machine and the automaton quite in the same sense in which Descartes had introduced them. Illustrating, for example, the force of habit, he said,

> We are automata as much as mind. . . . Our strongest and the most generally believed proofs are created by habit; habit directs the automaton which contains mind without thinking of it. . . . It is not enough if belief is compelled by conviction while the machine [l'automate] tends to believe the contrary. Both parts of our nature must be made to believe: the mind by reasons which it suffices to have seen once in a lifetime; the machine by habit and by not allowing it to incline to the contrary. (No. 241)

While thus accepting the automaton analogy for the composition of the human body, he rejected the complete identification of animals with machines:

> The arithmetical machine achieves results which come nearer to thought than anything the beasts can do; but it does nothing to justify our claiming that it possesses will [volonté] such as the beasts possess. (No. 163)

Christiaan Huygens (1629–95), in many basic respects a follower of Descartes, also had distinct disagreements with the philosopher who had been his father's friend and whom he had known in childhood. Committed to empiricism and personally inclining to the concrete and specific rather than the abstract and general, he disliked Descartes's style of reasoning with its a priori arguments, imaginary models, and analogies.[68] While his pendulum clock caused a horological revolution, and while he devoted much effort to the design of his "automatum planetarium," a clockwork-driven planetarium, Huygens had no taste for metaphors and called Descartes's "opinions concerning the soul of beasts . . . a ridiculous paradox."[69]

Baruch Spinoza (1632–77), also not strictly a Cartesian, used the image of the automaton repeatedly, usually with negative connotations. People who are ignorant and uncomprehending, he judged, "one has to regard as automata that are utterly mindless."[70] And commenting on contemporary—that is, Cartesian—psychology, he remarked that "the ancients said that true science proceeds from the cause to the effect; they have never conceived, as far as I know, as we have here, that the soul should act according to certain laws, like some spiritual automaton."[71]

Of Descartes's declared followers, it is sufficient here to sample only a few representative writings. The first major philosopher who unabashedly counted himself a Cartesian, the widely read and admired Father Nicolas Malebranche (1638–1717), was notably fond of the image of the watch. His best-known book, *De la recherche de la vérité* (1674–78), makes use of this image in a variety of ways. The design argument appears repeatedly in standard form, as for example in the following:

> When I see a *Watch,* I have reason to conclude, that there is some Intelligent Being, since it is Impossible for Chance and Hap-hazard to produce, to range and posture all its Wheels. How then could it be possible, that Chance, and a confus'd Jumble of Atoms, should be capable of ranging in all Men and Animals, such abundance of different secret Springs and Engines, with that Exactness and Proportion, . . .[72]

Allusions to the design argument also appear in various looser formulations; in one example, he observed that

> the transmutation of an Egg into a Chicken is infinitely harder than the bare Conservation of the Chicken when compleatly form'd: For as greater Art is requir'd to fabrick a Watch out of a piece of Iron, than to make it go when 'tis perfectly made.[73]

In other examples, clock imagery served to praise the complexity of creation[74] or to illustrate the different degrees of perfection in God's creation.[75]

Some of Malebranche's mechanical metaphors show the influence of Descartes. He reasserted, for example, the automatism of animals,[76] and he elaborated Descartes's theory of the soul.[77] And whereas animals were no more, perhaps, than mere automata, the same thing could occasionally be said even about humans.[78] Watches, which were of the same rank as clocks and automata, served to illustrate at length the two modes of existence—material and ideal—of dualist philosophy.[79]

Similar imagery appeared in the book *Traité de l'existence et des attributs de Dieu* (1712) by Bishop Fénélon, a prominent theologian and churchman of the later reign of Louis XIV; it was intended as a response to and elaboration of Malebranche's *Recherche de la vérité.*[80] Consistent with its title, the argument from design figured prominently, was presented in a number of variations, and was always illustrated by clock and machine imagery.[81] As Fénélon reiterated this theme in ever-new variations, he revealed a very distinct conception of the way in which things were to function.[82] In describing the world as a vast, complex interlocking system, he emphasized two peculiarities. First, each of its parts is indispensable for the stability of the system; if even the smallest element would fail to perform its prescribed function, the whole system would collapse. Second, the system has a centralistic command structure; the original design, continuing functioning, and ultimate survival of the whole system depend utterly upon a single authority.

Equivalent to the command structure of the world was that of the human body.[83] The authority of the human soul over its body was not only just as absolute but also as centralized as that of God over the universe.[84] As Fénélon described the programlike functions of the memory, the similarity between the abstract system that he was trying to outline and a mechanical automaton became palpable. There was, of course, another model of the same kind of system directly before Fénélon's eyes which would have been hard to ignore: the absolute state of Louis XIV.

The clock metaphor was now being used in a wide range of ways, many routine and trivial, some original and imaginative. Certain distinct topics began to stand out. One was the argument from design for which, despite much variation in detail, the underlying theme remained largely constant. Another was the notion of the automatism of animals. The more the controversies over this notion engaged the intellectuals, especially in the theological literature, the more the original mechanical analogy lost its freshness and receded into the background. These controversies, therefore, will not be considered further, since the returns from their study for the understanding of our present subject would diminish quickly, and since this subject has been adequately dealt with in a special literature.[85]

In the last decades of the seventeenth century, the mechanical, that is, Cartesian world view was becoming widely accepted among the educated public. Important in promoting such acceptance was a book by Bernard de Fontenelle (1657–1757), a versatile man-of-letters and savant, best known for the numerous *éloges* of scientists that he wrote during his long service as *secrétaire perpétuel* of the French Academy of Science. His *Entretiens sur la pluralité des mondes* (1686) was a popular but serious introduction into the world picture of the new science, presented from a Cartesian point of view and written in the form of a dialogue between a philosopher (the narrator) and an intelligent aristocratic lady. Characteristic is a passage from this work characterizing the mechanical nature of the new philosophy:

> I perceive, *said the Countess,* Philosophy is now become very mechanical. So mechanical, *said I,* that I fear we shall quickly be ashamed of it; they will have the World to be in great, what a Watch is in little; which is very regular, and depends only upon the just disposing of the several parts of the movement. . . . I am not of their opinion [who have less esteem of the Universe since they pretended to know it], *said she,* I value it the more since I know it resembles a Watch, and the whole order of Nature the more plain and easy it is, to me it appears the more admirable.[86]

This exchange confirms the central function of the clock metaphor in defining the mechanical philosophy. It also offers a glimpse into the emotional appeal of the new movement, its sense of departure into high adventure, and its confidence of "owning" the future. Fontenelle's book was successful and influential. Its form, the dialogue between philosopher and lady, became a favorite vehicle for popularizing science and was imitated by a number of notable authors.

In his mature years, G. W. Leibniz (1646–1716) spoke of the mechanical philosophy as a phase through which he had passed in his youth.[87] His writings are full of protestations of his own nonmechanical outlook and of expressions of disagreement with mechanical philosophers. At the same time, however, he made such frequent and continuing use of mechanical imagery and mechanical intellectual models—to say nothing of his activities as an inventor of actual machinery—that one suspects he never left mechanicism as far behind as he liked to think.[88]

When one surveys the clock and automaton metaphors in Leibniz's writing, one can observe a common characteristic: they illustrate processes that are deterministic. A particular expression of this bias was his doctrine of the "preestablished harmony" which, although applicable to all aspects of the creation, was designed specifically to clarify the relationship between body and soul.

Leibniz, like the Cartesians, insisted that body and soul were strictly separate and distinct and denied any interaction between the two. How then, was the complete congruity of their actions and experiences to be explained? Leibniz asserted that both body and soul were programmed ("preformed") at their creation, like automata, to such perfection that their actions would always be coordinated in perfect harmony.[89] He illustrated this scheme of action with the image of two orchestras playing the same music in perfect harmony but having no connection with each other,[90] and by the analogy of two clocks whose running is completely synchronous. Of this clock analogy, which was developed in considerable detail, the following are some salient phrases:

> My new hypothesis on the great question of *the union of soul and body* . . . may be made intelligible . . . by the following illustration. Suppose two clocks or two watches which perfectly keep time together [*s'accordent*]. Now that may happen *in three ways.* The first way consists in the mutual influence of each clock upon the other; the second, in the care of a man who looks after them; the third, in their own accuracy. . . . [There follow descriptions of the first and the second method.] Finally, *the third way* will be to make the two clocks [*pendules*] at first with such skill and accuracy that we can be sure that they will always afterwards keep time together. This is the way of pre-established agreement [*consentement*].
>
> Now put the soul and the body in place of the two clocks. Their agreement [*accord*] or sympathy will also arise in one of these three ways. . . . [He first refutes the theories of scholastic philosophy and of the Occasionalist brand of Cartesianism.] . . . Thus there remains only my hypothesis, that is to say, *the way of the harmony pre-established* by a contrivance of the Divine foresight, which has from the beginning formed each of these substances in so perfect, so regular and accurate a manner that by merely following its own laws which were given to it when it came into being, each substance is yet in harmony with the other, just as if there were a mutual influence between them, or as if God were continually putting His hand upon them, in addition to His general support [*concurrence*].[91]

Body and soul, then, are two automata programmed separately and independently but nevertheless working together in such perfect coordination that they seem to form a single unit. Leibniz reiterated this view on many occasions. In "The New System" (1695), he suggested that this preestablished harmony existed not only in the unity of the human body and soul but universely between the spirits and substances of the universe: "This mutual relationship, prearranged in each substance in the universe, produces what we call their *communication* and alone constitutes the *union of soul and body.*" This harmony was caused by automatonlike behavior of the participants: "For why might not God in the beginning give to substance an inner nature or force which could regularly reproduce in it—as in an *automaton* that is *spiritual or endowed with a living principle* . . . —everything that will happen to it?[92]

The theme returned repeatedly in the *Theodicy* (1710), Leibniz's attempt to reconcile the dogma of divine justice with the undeniable existence of evil in this world. In the preface, he summarized his earlier teachings.[93] To objections against the extreme determinism of his position, he offered this defense: "I was surprised to see that limits were placed on the power of God . . . even men often produce through automata something like the movements that come from reason."[94]

Such determinism clearly excluded any freedom of will: "As for *volition* itself, to say that it is an object of free will is incorrect. We will to act, strictly speaking, and we do not will to will." Our choices, in other words, are always predetermined: "All is therefore certain and determined beforehand in man, as everywhere else, and the human soul is a kind of *spiritual automaton.*"[95]

Nevertheless, Leibniz claimed, "nothing can be more favourable to freedom than that system." An automaton illustrated his argument:

> Just as if he who knows all that I shall order a servant to do the whole day long on the morrow made an automaton entirely resembling this servant, to carry out tomorrow at the right moment all that I should order; and yet that would not prevent me from ordering freely all that I should please, although the action of the automaton . . . would not be in the least free.[96]

Freedom, in short, was never more than the illusion of freedom. Leibniz later restated this deterministic scheme.[97]

The determinist metaphysics of Leibniz's preestablished harmony was elaborated further in his *Monadology* (1714). There he proposed that the universe, both in its material and spiritual aspects, was made up of a vast number of small force centers, particles resembling atoms or mathematical infinitesimals, which he called *monads*. The monads filled all spaces, everything was connected, and all transitions and graduations were continuous, without gaps. Monads were ordered in a pyramidlike hierarchy, with God as the supreme monad.

The monads did not interact. They were enclosed, "without windows," and were self-sufficient. Phenomena that gave the appearance of interaction between various material or spiritual agents were actually manifestations of the preestablished harmony. The various participating monads were preprogrammed separately for their entire career; their behavior was controlled only by this internal program and not by the contingent behavior of other monads. If, to all appearances, a man was killed by a rifle shot, the actions of his bodily and spiritual monads were indeed independent of those of the monads of the bullet; they were merely perfectly coordinated.

The determinism postulated here was that of an automaton; the automaton analogy that Leibniz had invoked in earlier writings was to serve again in *The Monadology.* He not only applied it to the living organism as a whole;[98] the monads themselves, of which the whole organism was composed, were also comparable to automata.[99]

Leibniz wrote *The Monadology* in 1714. In the two remaining years of his life, the clockwork metaphor occupied him once more on another occasion. In 1715, a controversy with Sir Isaac Newton, in which Leibniz had been involved since the 1690s, entered a new stage. This time, Leibniz's opponent was not Newton himself but a younger disciple close to the master, the Reverend Samuel Clarke. Their subject was the attributes and characteristics of God. In the question, "Is God's chief characteristic unlimited *will* or unlimited *intellect?*" Clarke defended the voluntarist position and Leibniz the intellectualist, or determinist position. The clock metaphor once again played a prominent role. The debate was terminated only by Leibniz's death. Published promptly thereafter in the form of a dialogue of letters, it attracted a large audience.[100] While both sides, Newtonians and Leibnizians, claimed victory, the effect of the debate upon the clock metaphor was clear. In the first two exchanges, Clarke inflicted so much damage to it that Leibniz stopped using it. In hindsight, the debate appears to have been a turning point in the fortunes of the clock metaphor (for a full discussion of the controversy, see pp. 98–101 below).

Leibniz's immediate followers, however, managed to ignore the setback. Indeed, in the work of Christian Wolff, the clock metaphor experienced a late flowering that has not been matched in its luxuriousness.

Christian Wolff's misleading reputation as a trivializer of Leibniz's philosophy has obscured his real significance as the originator of that procedural orderliness and methodological rigor that has characterized German philosophy ever since. The most important as well as most typical portion of his output is a series of books written in German (rather than in the customary Latin) during his first term as professor of philosophy in Halle (to 1723) which treat, one by one, all the traditional branches of philosophy (logic, metaphysics, ethics, and so forth). Wolff's treatment of metaphysics (known as the "German Metaphysics," to distinguish it from a later Latin version) is divided into ontology, pneumatics (*scientia spiritum,* i.e., psychology and natural theology), and cosmology, the theory of the world.[101] It is the cosmology that is of interest here, for it is formally and comprehensively based upon the analogy of the clock. In no other serious work of philosophy, before or since, has the clock image played so important a role.

Wolff began with an elaborate demonstration, step by step, that the world is a machine. First he defined the world as a series of unchangeable things which exist in continuity in space and time and which are interconnected in their totality so that they form a single whole.[102]

The character of any given world is fixed by the particular composition of its parts and is thus unchangeable. More than one world is conceivable; clocks, after all, are made in various types, with wide differences in the number, shape, and arrangement of their internal parts. But once the individual character of a complex artifice is fixed by its particular composition, Wolff concluded, its career through time is fixed as well. "And thus the things in this world are linked together in time because they are linked together in space."[103] To reinforce this point, Wolff added another clock analogy: "The world functions no differently than a clockwork. For the essential character of the world is rooted in its manner of composition, and so is the character of a clock."[104]

Wolff asked the reader not to be put off by his selection of clockwork and machinery for his metaphors, "for the world likewise is a machine. The proof," he went on, "is not hard. A machine is a composite artifice whose movements are determined by their manner of composition. The world is likewise a composite thing whose changes are determined by its manner of composition. Therefore the world is a machine" (§557).

Concerning the question of how the world possessed order and therefore truth, he concluded, "the reason that there is truth in the world is that it is a machine. Were it not to remain a machine, then the distinction between it and a dream would vanish" (§559). Consequently, all composite things, the world included, were machines and, just for that reason, contained truth.[105]

Finally he returned to the determinist character of the world: "Since the present state of the world is rooted in the preceding one, and the future state in the present one; thus the events in the world obtain their certainty [Gewissheit], and thus by the fact that the world is a machine all events therein are rendered certain" (§561). The world, in short, is a machine, and what the world and machines have in common is the rigid, preprogrammed determinism of their actions in time.

Such was the general program of Wolff's cosmology, as outlined in the first few pages of the cosmological chapter of his "German Metaphysics." In subsequent paragraphs he elaborated upon it at great length, partly going over points already established, partly developing subsidiary arguments. In all this, the clock metaphor was ubiquitous. He offered at least a dozen fully developed clock analogies,[106] usually introduced formally with phrases such as "I shall elucidate this by means of the simile of the clock" or "as the example of the clock shows," not counting the cursory references to clockwork that are far more numerous. In virtually all instances, the clock metaphor supported a view of the universe that was rigidly deterministic.

To give an impression of the relentlessly repetitive thoroughness with which Wolff developed his system and of the clock metaphor's predominant role in it, the remainder of his argument will be outlined briefly here.

According to Wolff, the course of the world is predetermined for all times from the moment of its creation. If, for example, we have this morning a clear sky and a brisk wind, this has its reasons for being in the particular way in which our world has been created. Should it have rained instead, the world would have had to be created differently. "To those who have trouble understanding this, I again present the example of the clock," Wolff explained, before launching into another lengthy clock analogy.[107] This particular argument went on for several chapters.

The determinism outlined here was elaborated at great length in all its particular aspects, always with the help of clock analogies. If, for example, the world's particular construction fixed its characteristics and hence its fate through time, the smallest modification of any of its parts would affect these things.[108] Or if the mechanism's predetermined course was to be altered, this could be done only by outside agents. In the case of the clock, this was equivalent to human interventions such as winding or setting; in the case of the world, such a change was conceivable only in the form of actions by a deity overruling the laws of nature, in other words, by miracles. The character of miracles was clarified further with the help of the clock image.[109]

Wolff went on to explain that, for both the clock and the world, outside intervention would forever alter their subsequent careers; the only way of cancelling out the effects of a miracle was another miracle.

The essential characteristic of a miracle is that its author is located independently and outside of the world in which the miracle is being observed; it is "the same as the moving of a clock hand from outside" (§640). What does Wolff mean by "outside of the world"? Things are "outside of the world" if their existence does not depend on the world, "just as I say that a man who shifts the hand of the clock is outside of the clock, because his existence does not depend on the clock nor does he, unlike its parts, belong to the clock" (§641).

Even a radical determinist like Wolff had to leave some freedom of action for God. He accomplished that by making a distinction between "certainty" and "necessity." Although the career of the world, like the running of a clock, was predetermined or certain (gewiss), it was not inevitable or necessary (notwendig), for its creation had been accidental (zufällig) or, in the preferred technical term, *contingent.*[110] Different kinds of worlds might have been possible; after all, clocks can be made in many different types, differing in the number, shape, and arrangement of their internal parts. God had the freedom, then, to choose the particular form of his creation.

Contemporaries saw Wolff as an extreme determinist. When enemies suggested to his sovereign, the irascible "soldier king" Friedrich Wilhelm I of Prussia, that Wolff's teaching implied that soldiers were automata and that punishing deserters was useless and irrelevant, the king in 1723 removed Wolff from his professorship and banished him from town and country.[111] In his exile in Marburg, Wolff began to write a new version of his encyclopedic treatment of philosophy, this time in Latin. Cosmology, previously subsumed under metaphysics, was now treated independently in a *Cosmologia Generalis* published in 1731, which largely covered the same ground as Wolff's earlier work.[112] His commitment to the analogies of machine and clock was undiminished. Key paragraphs had provocative headings such as "Mundus omnis, etiam adspectabilis, machina est" (the whole and visible world is a machine, §73); "Omne ens compositum machina est" (every composite thing is a machine, §74); and "Mundus propemodum se habet ut horologium automaton" (the world functions like a clockwork automaton, §117). And the equivalence of world and clock was argued at length (§§117 and 118) in much the same manner as in the earlier *German Metaphysics.*

Concerning determinism, however, Wolff had somewhat tempered his position. While continuing to maintain that the character and, barring outside intervention, the career of the clockwork world were fully determined by the original construction of the mechanism, he took care to explain that absolute necessity ("necessitas absoluta") was equally alien to both the world and the clock because of the contingent character of their origins. The discussion culminated in the statement, "If the only common characteristic is a predetermined manner of functioning in contrast to the freedom of man, then the world is like a clock" ("Tertium comparationis minime est modus agendi necessarius in oppositione ad libertatem hominis expensus, dum mundus cum horologio comparatur").[113] What the two have in common, then, is a complete lack of that freedom of will and choice which is the distinguishing mark of man.

Wolff's cosmology no doubt constitutes the most comprehensive and methodical attempt to make the clockwork metaphor, or indeed analogy, the foundation of a philosophic system.[114] The reason Wolff chose the clock for this distinguished task was his belief in determinism as the basic characteristic of the world, a characteristic, he thought, that both the world and the clock had in common. This belief, however, as Wolff learned from personal experience, was becoming controversial. It contradicted the principles of free will and freedom in general, which a steadily growing portion of the public was inclined to prefer. For the rest of the eighteenth century, the clockwork analogy was discussed in German philosophy almost exclusively from the point of view of the debate over determinism versus freedom of will, with an increasing partiality for the latter (see chapter 6).

The later career of the clock metaphor in French philosophy still remains to be considered. Having been employed there earlier without the conspicuous enthusiasm that marked its use in the British (briefly) and German literature, the metaphor survived in France a good deal longer. In Britain, the popularity of the metaphor, limited as it was, culminated in the writings of Robert Boyle and in Germany in those of Christian Wolff, beginning its decline, in each case, soon afterward. If a somewhat tentative survey of eighteenth-century writings on natural and political philosophy is reliable, it never became controversial in France. It remained in use in its conventional form, without special emphasis and without the enthusiasm that is aroused by anything novel. When the frequency of the metaphor in French philosophical writing decreased toward the end of the eighteenth century, it was probably more because it was considered tiresome than wrong.

Voltaire's (1694–1778) uses of the clock metaphor were conventional. In the majority of cases, he linked it with the design argument. In his *Traité de métaphysique* (1734), he argued for the existence of God not only from the order but also from the purposefulness of the design of the universe.[115] Along with this affirmation of final causes, he asserted his opposition to Descartes and Leibniz and his solidarity with Newton; in support of the same position, he invoked Newton's authority again in his *Éléments de la philosophie de Newton* of 1738.[116] Other uses of the metaphor by Voltaire in connection with the design arguments are found in later works, for example, in his *Dictionnaire philosophique* (1751) and in the satire "Les Cabales" of 1772.[117]

Voltaire's frequent and comfortable use of clock imagery does not establish him as a mechanist. When comparing the soul with "the spring of our clockwork," for example, he was not trying to explain the soul mechanically.[118] More likely, he had never given the clock metaphor much thought; if it was more than a conventional figure of speech to him, it embodied merely a useful mechanism of ingenuity and complexity which ordinary mortals could not and need not understand.[119]

The attitude of Denis Diderot, another philosophe whose life spanned most of the eighteenth century, toward mechanism, machinery, and clockwork was similar to Voltaire's. He used mechanical imagery freely and frequently but in a loose and casual manner. Liberty appealed to him and determinism made him vaguely uncomfortable, but when forced to choose he would come down on the side of determinism.[120]

With regard to the creation, animate or inanimate, he held conventionally mechanistic views. In endorsing the design argument, he stated that the world "is simply a machine, with its wheels, its ropes, its pulleys, its springs, and its weights."[121] Living beings, he thought, were mechanisms of the same kind. This is implied in his *Encyclopédie* entry for "méchanisme," which he defines as "the manner in which some mechanical thing produces its effect; thus one speaks of the *mechanism* of a watch, the *mechanism* of the human body."[122] Discussing mechanistic medicine, a doctrine of medicine that he supported, Diderot said that, according to that theory, "the animal body, and consequently the human body, is considered a real machine."[123] In an attempt to explain the processes of human sense perception to persons incapable of abstract thought, he constructed an elaborate clock analogy:

> Consider man, the automaton, as a walking clock where the heart represents the main spring, and the parts contained in his chest the other principal pieces of the movement. Imagine in his head a bell equipped with little hammers linked to an infinite number of strings that end at all points of the body. Mount on the bell one of those little figures with which we adorn the top of our pendulum clocks; it has the ear bent forward like a musician listening if his instrument is well tuned: this little figure be the *soul*.[124]

From this analogy, he developed an entire theory of perception in terms of a central bell and multiple hammers activated by strings from the various parts of the body. Noteworthy was Diderot's continued struggle with the problem of the soul; what Descartes had earnestly and literally identified with the pineal gland was now lightheartedly illustrated by an automated puppet.

The same problem occasioned another clock metaphor in Diderot's later *Elements of Physiology,* where he expressed an unmistakable scepticism toward the notion of soul in general:

> Animal and Machine.—What difference between a watch that feels and lives, and a watch of gold, iron, silver and copper? If a soul was attached to the latter, what would it accomplish there? If the union of a soul with this machine is impossible, let someone prove it to me. If it is possible, let

> someone tell me what the effects of that union might be. The peasant who sees a watch move and who, unable to understand the mechanism, puts a spirit into the hand, is neither more nor less foolish than our spiritualists.[125]

With the suggestion that the concept of soul is only a disguise for ignorance, Diderot came close to contemporary materialists like La Mettrie and Holbach who deny the existence of a spiritual world, that is, the possibility of the soul as well as of God. The shift from mechanism to materialism is subtle and seems, on the surface, to involve little change. Curiously, however, the materialists seem to have had comparatively less interest in the clock metaphor. La Mettrie's notorious *L'Homme machine,* where "machine" certainly meant "clock," makes some explicit statements to that effect ("The human body is a watch, a large watch constructed with such skill and ingenuity"),[126] but they are few and incidental. When he claims that "the body is but a watch, whose watchmaker is the new chyle [*sic*]" (p. 186), the purpose of the metaphor seems to be only to parody the design argument, where the watchmaker had been God. There are a few more implicit allusions to the clock, as in "The human body is a machine which winds its own springs. It is the living image of perpetual motion" (p. 154), and "the heart is the mainspring of the machine" (p. 190). But on the whole, the book projects La Mettrie's conviction that the clock analogy has been evoked often enough, that one may take it for granted, and that it would be tiresome to belabor it more.

How intimately the clock image was connected with the French Enlightenment's conception of order emerges in the remarkable entries for "system" and "harmony" in Diderot's *Encyclopédie* (the question of their authorship, which is uncertain, is not crucial; if not written by Diderot himself, they must have met his and his coeditor's approval). Both entries deal with the rationally ordered integration of the parts into a complex whole, and both illustrate their arguments with clock imagery.

The entry for system begins:

> *System* is nothing more than the disposition of the different parts of an art or a science into a state where they all mutually support each other and where the last ones are explained through the first. Those that account for the others are called *principles,* and the *system* is all the more perfect as the principles are fewer in number: indeed, it is desirable that they should be reduced to a single one. For, just as there is one main spring in a clock upon which all others depend, there is also in all *systems* one first principle to which the different parts that make it up are subordinated.[127]

The clock here is more than mere illustration; its characteristic logical structure seems to have suggested to, or at least reinforced in, the author a preference for a centralist, hierarchical organization where all functions are subordinated to and derived from one single authority.

Equally revealing is the role of clock imagery in the definition of the concept of "harmony." Defined as "the general order that prevails between the various parts of a whole" owing to which, these parts "work together as perfectly as possible," it was a quality that, to some extent, existed in the eye of the beholder. To appreciate harmony, it was necessary to appreciate function, extent, and complexity of the total system in which it prevailed. To illustrate the varying levels of depth of such appreciation, the author introduced the watch:

> If a peasant got a watch into his hands for the first time, he would look at it and he would perceive some coordination between its parts; he would conclude from this that it had its use; but that use being unknown to him, he would have gone no further or he would have been wrong. Let us put the same machine into the hands of someone more informed or more intelligent who discovers in the uniform motion of the hand and in the equal divisions of the dial that it might well have been designed to measure time: his admiration would be greater. The admiration, however, would be much greater yet if the examining mechanic were able to give an account of the disposition of parts in relation to the known effect not only to himself but also to others to whom the watch was presented for examination. The more complex a machine, the less we are able to judge it.[128]

The moral of this story was, of course, to warn us not to presume to make judgments about the harmony of systems that we are not qualified to judge or, worse yet, to take practical actions on the basis of such unsound judgments. In conclusion, the article listed some typical systems to which the term *harmony* was commonly applied, a list that is impressive in the importance of the examples given and the closeness to which it places the clock image with some of the crucial issues of the period:

> The word *harmony* had been applied to the *art of governing* and it is said, "He commands great harmony in that state; to the arts and their productions, but above all to the arts whose purpose is the use of sounds and colours. Also, one has spoken of the *general harmony of things, the harmony of the universe.* See World, Nature, Optimism, etc.

The article on harmony in Diderot's *Encyclopédie* summarizes the status of the clock image in France in the second half of the eighteenth century. Formulated and read by the more progressive intellects of the country, the article restated the centuries-old affinity between the clock and one of the period's highest values—a value characterizing perfection in man's most important concerns, from the Divine Creation to human art. Another Frenchman, Nicole Oresme, bishop of Lisieux, had expressed much the same thing almost four centuries earlier.

British Mechanists

In Britain the clock metaphor flourished best in the second half of the seventeenth century or, more exactly, from the time when the works of Descartes first received wide attention to the broadly publicized debate between Leibniz and Samuel Clarke in 1715–16. This flowering of the clock metaphor, admittedly a phenomenon of miniature dimensions then, roughly coincided with a series of revolutionary changes of major importance. The clock metaphor, it seems, became popular in the last years of the English Civil War, culminated during the Restoration, began its decline after the Glorious Revolution, and was put to rest when the installation of the Hanoverian kings cemented the permanent establishment of a constitutional monarchy. The same time span also covered the triumph of the Scientific Revolution, from the founding of the Royal Society to the Newtonian synthesis, and the conclusion and final settlement of the religious struggles that had been set in motion by the Reformation. These social, political, scientific, and religious revolutions did not occur separately and independently but were interwoven in their causes and dynamically interacting in their consequences. This is mirrored on a small scale in the story of the clock metaphor in Britain, which, too, involves discussions of scientific, theological, and politico-social subject matter.

The revolutions of late seventeenth-century Britain were comparatively nonviolent. Their battles were conducted not so much with sword and gun as with pen and printing press, in a vast amount of polemics and debates on all levels of formality, ranging from verbal discussions, letters, and pamphlets to learned books. To do justice to the fortunes of the clock metaphor, it would be necessary to sift through the full mass of such materials and interpret them against the background of the several major and countless minor debates. Here, for the time being, it is only possible to present those fortunes as though they formed a separate coherent development and to hope that this procedure will not do too much violence to historical reality.

The clock metaphor became popular in British natural philosophy, it seems, through the writings of Descartes. The question about extent and significance of Descartes's influence on British philosophy in the seventeenth century is the subject of a sizable literature and a continuing debate.[129] This much seems safe to say: from the late 1640s, British philosophers were aware of and familiar with Descartes's writings and participated in the debates about the Cartesian doctrines. Descartes's teachings were perceived to be in direct contrast to the teachings of Francis Bacon, which had dominated all scientific and philosophical activities on the island. Apart from a few early avowed Cartesians like Sir Kenelm Digby and Henry More, most British writers reacted to Descartes with caution and reserve. They tended to stress the aspects in which they differed and to avoid the appearance of being under the Frenchman's influence. In time, however, more and more British thinkers recorded their admiration and at least partial approval of Descartes, and as it became clear that his philosophy proved a valuable supplement to that of Bacon, much of it was quietly accepted.

The Royal Society's campaign for sobriety of expression and against ornateness of style evidently was unable to suppress the use of the clock metaphor. Robert Boyle (1627–91), the most influential English philosopher between Bacon and Newton, used the metaphor with gusto. "To explain this a little, let us resume the often mentioned, and often to be mentioned instance of the clock," runs the beginning of a typical passage.[130] Like many of his contemporaries, he was obviously charmed by this machine; he repeatedly expressed his admiration for the Strasbourg Cathedral clock and carried clock imagery even into writings that were poetic and devotional and entirely nonscientific.[131]

For all his personal fondness of clocks, however, he used this metaphor and others in general judiciously and for carefully considered purposes. He justified this habit, perhaps with an eye on Thomas Sprat's injunction, in the preface to his last major work (1690): "In the following treatise, as well as in divers of my other writings . . . , I make frequent use of similitudes, or comparisons: and therefore I think myself here obliged to acknowledge once and for all, that I did it purposely."[132] He insisted that metaphors helped to clarify matters, lent strength to arguments, and helped to avoid "frequent and tedious circumlocutions."[133] In further defense of the habit, he invoked the names of "the illustrious Verulam" and of "that severe philosopher, Monsieur Des Cartes himself, [who] somewhere says, that he scarce thought, he understood anything in physics, but what he could declare by some apt similitude," to which Boyle added that "proper comparisons do the imagination almost as much service as microscopes do the eye."[134] This, it seems, expressed the views of many.

Clockwork metaphors in British seventeenth-century natural philosophy concentrated on certain favorite themes: the new approach to solving the riddles of nature, the characteristics of the mechanical philosophy, mechanical aspects of natural phenomena, and the relationship between nature and its Creator. While the partisans of such a broadly based and many-featured movement as the new science were bound to have diverging ideas on its definition and goals, they had little difficulty agreeing on what they were against. Their principal enemy was the Aristotelian scholasticism that had dominated the teaching and practice of philosophy for centuries. The clock image illustrated the contrast between the old and the new philosophies. Simon Patrick, a liberal clergyman and later bishop of Ely, illustrated the futility of the scholastic approach in a parable (1662) of a fraudulant clockmaker who tried to browbeat his clients by his mastery of Aristotelian jargon. He would expound impressively on "whether the hammer were the intelligencer of the Bell, and consequently whether *forma informans* or *assistens;* whether the Bell did act upon the elementary parts of the clock, or they upon the Bell; . . ." and so forth[135] until he was exposed by a man trained in modern science who took the faulty clock apart and identified the trouble in the worn teeth of a wheel.

Others treated the theme similarly. The striving for a full understanding of "the management of this great Machine of the World," Henry Power declared in 1664,

> . . . must needs be the proper Office of onely the Experimental and Mechanical Philosopher. For the old Dogmatists and Notional Speculators, that onely gaz'd at the visible effects and last Resultances of things, understood no more of nature, than a rude Countrey-fellow does of the Internal Fabrick of a Watch, that onely sees the Index and Horary Circle, and perchance hears the Clock and Alarum strike in it: But he that will give a satisfactory Account of those *Phaenomena,* must be an Artificer indeed, and one well skill'd in the Wheelwork and Internal Contrivance of such Anatomical Engines.[136]

Matthew Hale illustrated this contrast between verbal speculations and empirical knowledge in the various theories of the creation of the world by a clock anecdote: spokesmen of the various philosophical schools—Epicureans, Aristotelians, and so forth—are assembled before a clock, competing to explain it in the jargons of their philosophies, then a clockmaker arrives and exposes all their speculations as pure nonsense. The situation was the same, concluded Hale, with the "*Hypotheses* of the Learned Philosophers in relation to the Origination of the World and Man."[137]

Boyle, too, found clock imagery effective in exposing the speciousness of scholasticism. All that these "pretended explanations" of the "school-philosophy" were capable of doing was attributing the "wonderful properties of the load-stone, to the attractive faculty of amber . . . , and the medical virtues of gems [to] what they call their substantial forms." The clock image demonstrated the absurdity of such thinking:

> He must be a very dull inquirer, who, demanding an account of the phaenomena of a watch, shall rest satisfied with being told, that it is an engine

> made by a watch-maker; though nothing be thereby declared of the structure and co-aptation of the springs, wheels, balance, and other parts of the engine, and the manner, how they act on one another.[138]

The purpose of the clock metaphor, however, was by no means only to discountenance scholasticism; it also performed a constructive and much more important function, namely, to define an essential ingredient of the new mechanical philosophy. Its method of uncovering the secrets of nature was that of the clockmaker; to find out how an unfamiliar clock worked or why a broken clock did not, the clockmaker would take it apart. The taking apart of a clockwork became an illustration of that process known as *analysis*.

Joseph Glanvill, a stalwart of the early Royal Society and a mechanist despite close ties to the Cambridge Platonists, warned in 1665 that we would never understand nature if we only observed her dial plate and hour hand but failed to "see the first springs and wheeles that set the rest a going."[139] Instead, he implied, we must take the phenomena apart like a clock. " 'Twere next to impossible for one, who never saw the inward wheels and motions, to make a watch upon the bare view of the *Circle of hours,* and *Index*. . . . Yea, the most common *Phaenomena* can be neither known, nor improved, without insight into the more *hidden* frame."[140] Elsewhere again (1668), when describing the benefits of chemistry, notably "these *useful* and *luciferous processes,* by which Nature is *unwound,* and resolv'd into the *minute Rudiments* of its *Composition,*" he compared chemical analysis to the disassembly of a clock: "We cannot understand the *frame* of a *watch,* without taking it into pieces; so neither can Nature be well *known,* without a resolution of it into its *beginnings,* which certainly may be best of all done by *Chymicial Methods.*"[141]

The thought was expressed in the same manner but even more forcefully by the physician Thomas Willis, discussing the complexities of muscular action in 1684.[142] And Roger Cotes, in his preface to the second edition (1713) of Newton's *Principia,* ridiculed those who would rather idly speculate than investigate empirically: "But if a certain clock should be really moved with a weight, we should laugh at a man that would suppose it moved by a spring . . . he ought to . . . look into the inward parts of the machine."[143]

The thoroughness of understanding that a mechanic was expected to have of a clock, then, became a paradigm for positive knowledge. We genuinely understand any process, said Thomas Sprat, "of whose motions of all sorts, there may be as certain as accompt given as of those of a Watch or Clock."[144] John Locke, in the belief that the ultimate test of knowledge is the ability to predict, declared: "Did we know the mechanical affections of the particles of [some living organism] as a watchmaker does those of a watch, whereby it performs its operations . . . we should be able to tell beforehand" its characteristics.[145] And Robert Hooke, trying to phrase a strong endorsement for the newly invented microscope, predicted that this instrument would lead "towards the increase of the *Operative,* and the *Mechanick* Knowledge, to which this Age seems so much inclined" by enabling us "to discern all the secret workings of Nature, almost in the same manner as we do those that are the productions of Art, and are manag'd by Wheels, and Engines, and Springs" or, in other words, by clockwork.[146]

As a master clockmaker's familiarity with the inner workings of a clock was a symbol of the certainty of empirical knowledge, conversely, the doomed attempts of a layman to achieve an understanding of clockwork while confined to observing it from the outside illustrated the ambiguities of theories based on speculation and hypothesis. For some writers, for example, Henry Power and Joseph Glanvill, the contrast between philosophizing like an outside observer or like an expert clockmaker was the difference between the old and the new or, indeed, the wrong and the right approaches to knowledge about nature.

Others did not share such unrestricted self-assurance. While working with the theme of the unquestioned superiority of the clockmaker's understanding of clockwork over that of the outside observer, they sometimes admitted doubt that even the new science had unfailing access to inside knowledge and acknowledged that at times there was no choice but to resort to hypotheses. Such hypotheses, however, as Boyle urged, should be used with caution.

What "many atomists and other naturalists" confidently advance as "the true and genuine causes of the things they attempt to explicate," he observed, were often mere hypotheses, no better than explanations of a clock's movement given by someone who could see it only from the outside. A given phenomenon may be explained by various competing hypotheses; indeed, nature has available many different means to produce the same effect, just "as an artificer can set all the wheels of a clock going, as well with springs as with weights." What specific device, however, nature has actually employed in every particular case, man, with his "dim reason," may find "very difficult if not impossible" to determine. We must not think that an explanation is the correct one merely because it is within our modest intellectual grasp, for we should "consider how apt an ordinary man, that had never seen the inside but of one sort of watches, would be to think, that all those [other types of clocks and watches] are contrived after the same manner, as that, whose fabrick he has already taken notice of." Rather, we should keep in mind the experience of a man visiting "a skilful watchmaker's shop [who] shall observe, how many several ways watches and clocks may be contrived." Therefore, we should not be content with hypotheses that are merely possible but must seek positive information, that is, we must

> . . . declare what precise, and determinate figures, sizes and motions of atoms, will suffice to make out the proposed phaenomena, without incongruity to any others to be met with in nature: as it is one thing for a man ignorant of the mechanicks to make it plausible, that the motions of the famed clock at Strasburg, are performed by the means of certain wheels, springs, and weights, etc., and another to be able to describe distinctly the magnitude, figures, proportions, motions, and, in short, the whole contrivance . . . of that admirable engine.[147]

Although the superiority of empirical knowledge over hypothetical explanation was a generally accepted truism, the contrast between the two was sometimes held up as a reminder of human limitations and as a call for humility. In a sense, we all are only external onlookers upon the clockwork of nature, and whenever we succeed in catching a glimpse of its inner workings, we have God to thank for it. In his early *Usefulness of Natural Philosophy,* Boyle wrote:

> As highly as some naturalists are pleased to value their own knowledge, it can at best attain but to understand and applaud, not emulate the productions of God. For as a novice, when the curiousest watch the rarest artist can make, is taken in pieces and set before him, may easily enough discern the workmanship and contrivance of it to be excellent; but had he not been shewn it, could never have of himself devised so skilful and rare a piece of work.[148]

This thought was expressed again later in the same work[149] and was repeated again, in simpler formulation, by Thomas Sydenham and a little later by John Locke.[150]

Locke, who dealt with these problems extensively in his *Essay concerning Human Understanding* (1690) and whose conclusions were considerably more differentiated than those of his predecessors, continued to work with the traditional clockwork imagery. He assumed that the connections between the properties of the organisms and substances of nature and their physical constitution were of the same kind as those between the internal mechanisms of a clock and its external expressions, such as chiming, ticking, and turning. Since our chances are remote, for example, of obtaining an accurate insight into the mechanisms that relate the properties of chemical substances to their structures, our understanding of nature is bound to remain fragmentary and tentative. This, he maintained, is true for all parts of nature, starting with the human body:

> Had we such a knowledge of that constitution of man, from which his faculties of moving, sensation, and reasoning, and other powers flow . . .

> as . . . it is certain his Maker has, . . . our idea of any individual man would be as far different from what it is now, as is his who knows all the springs and wheels and other contrivances within the famous clock at Strasburg, from that which a gazing countryman has of it, who hereby sees the motion of the hand, and hears the clock strike, and observes only some of the outward appearances.[151]

Locke extended this line of reasoning to the limitations of our knowledge of nature in general, again using the same imagery.[152]

Locke's explanation of why our understanding is so limited resolved, at the same time, the contradiction between the two ways of viewing a clock:

> He that was sharp-sighted enough to see the configuration of the minute particles of the spring of a clock, and observe upon what peculiar structure and impulse its elastic motion depends, would no doubt discover something very admirable: but if eyes so framed could not view at once the hand, and the characters of the hour-plate, and thereby at a distance see what o'clock it was, their owner could not be much benefited by that acuteness; which, wilst it discovered the secret contrivance of the parts of the machine, made him lose its use.[153]

The superiority of the inside over the outside view is no longer maintained; both are required for an adequate picture of reality. Man, however, is not equipped for this comprehensive vision.

If the difference between the old and the new ways of studying nature was that between looking at a clock from the outside or taking apart its movement, and if a clockmaker's understanding of the functioning of clockwork rated as the paradigm of positive knowledge, then it is easy to see why the new philosophy was often called mechanical. But there were other ways of defining the mechanical philosophy, and there were other aspects in turn, to that larger and more complex movement of the new science than the mechanical. To get a glimpse of the role of the clock metaphor in the larger movement, it is instructive to return to the writings of Boyle, who was praised by contemporaries as "the introducer, or, at least, the great restorer of mechanical philosophy among us."[154] While Boyle disparaged all categorization, especially of himself, and while modern observers consider Boyle too eclectic simply to classify him as "mechanist," there is something to be said for the evaluation by his contemporaries. Boyle admired Descartes and was deeply influenced by him.[155] The adjective "mechanical," whatever its definition, meant to him a positive quality that stood in contrast to "Aristotelian" and "mystical."[156] His perhaps most important scientific contribution in the technical sense, his *corpuscular philosophy,* was a mechanistic theory of matter which tried to account for material properties in terms of the mechanical characteristics of the smallest particles, a contribution that has earned him the title of "father of modern chemistry."[157]

Boyle viewed the mechanical and corpuscular philosophies as largely the same thing, and he repeatedly called upon the clock metaphor to define it.[158] Emphasizing the simplicity and clarity of that philosophy, Boyle insisted "that there cannot be fewer principles than the two grand ones of mechanical philosophy, matter and motion."[159] Knowledge of the latter, namely, of the motion of all particles involved in physical phenomena, was indispensable, for without it "we can little better give an account of the phenomena of many bodies, by knowing what ingredients compose them, than we can explain the operations of a watch, by knowing of how many, and of what metals the balance, the wheels, the chain, and other parts are made."[160] Sometimes he employed clockwork imagery even more directly to define his philosophy:

> For [according to] that [corpuscularian] hypothesis, supposing the whole universe (the soul of man excepted) to be but a great Automaton, or self-moving engine, wherein all things are performed by the bare motion (or rest), the size, the shape, and the situation, or texture of the parts of the universal matter it consists of; all the phenomena result from those few principles, single or combined, (as the several tunes or chimes, that are rung on five bells,) and these fertile principles being already established by the inventors and promoters of the particularian hypothesis; . . . So that the world being but . . . a great piece of clock-work, the naturalist, as such, is but a mechanician; however the parts of the engine, he considers, be some of them much larger, and some of them much minuter, than those of clocks or watches.[161]

While a precise definition of the corpuscularian or mechanical philosophy would have stressed its commitment to accounting for natural phenomena through knowledge of "matter and motion" of the smallest particles involved, it was much easier and more economical to characterize the new philosophy through the similarity of its methods with those of the clockmaker and to refer to nature itself as a clockwork. Expressions to that effect are ubiquitous in Boyle's writing.[162]

He and his contemporaries applied the clock metaphor to a certain number of major themes and also used it routinely in innumerable casual phrases. They occasionally invoked it in fresh contexts in illustration of some particular observation. Boyle, for example, offered the essential similarity between a mighty tower clock and a dainty pocket watch as a demonstration that the laws of nature are valid on any scale, large or small.[163] Many spectacular natural phenomena, he observed, have causes that are surprisingly small,

> as in a clock, a small force applied to move the index to the figure of XII will make the hammer strike often and forcibly against the bell, and will make a far greater commotion among the wheels and weights, than a far greater force would do, if the texture and contrivance of the clock did not abundantly contribute to the production of so great an effect.[164]

Boyle seemed to regard this ability to produce a large effect by small means, which is frequently referred to elsewhere in his writings (although without explicit comparison with clockwork), as an essential characteristic of mechanics.[165]

John Locke questioned the absoluteness of systems of classification that the scholastic philosophy had been so fond of, with their hierarchical structures, by pointing to the dependence of their categories upon semantic accidents.[166] The true distinctions among classifiable concepts and objects and the principles for ordering them systematically, however, can be discovered only by analyzing unfamiliar materials in the manner of clockmakers. Unfortunately, that is not how we proceed in practice. Instead, our categories are defined and classified simply according to need and to the words that happen to be available in our language: "each abstract idea, with a name to it, makes a distinct species."[167]

Robert Hooke finally suggested an indirect, circumstantial method for obtaining knowledge about nature in cases when the method of direct analysis, in the manner of a clockmaker's taking apart a clock, was not applicable:

> There may also be a Possibility of discovering the Internal Motions and Actions of Bodies by the sound they make, who knows but that as in a Watch we may hear the beating of the Balance, and the running of the Wheels, and the striking of the Hammers, and the grating of the Teeth, and the Multitudes of other Noises."[168]

Clock metaphors such as the foregoing represent the high point of the mechanical philosophy in Britain. The metaphors and the themes they illustrated were uncontroversial, free of theological or political provocation, and accepted without reserve. Within decades, however, soon after the turn of the eighteenth century, the mechanical philosophy lost its popularity, and applications of clock metaphors became rarer, especially when used to express approval. This shift had to do with certain further uses of the metaphor that not only provoked religious sensitivities but also challenged some basic cultural values; one involved the concept of animal automatism, the other the "argument from design."

Since the early seventeenth century, there had been a gradual increase in the practice, in Britain as well as on the Continent, of comparing specific functions of a living body as well as entire animal organisms with mechanical processes, machines, and, inevitably, clockwork. Earlier we saw examples from philosophers and poets which, whether serious or playful, were hardly more than literary embellishments, without any intention of making a philosophical point (see chapter 2). This colloquial use of the clock metaphor increased in the course of the seventeenth century with the introduction of mechanical methods into medical practice and with the appearance of the landmark works of Harvey, Descartes, and Borelli. Henry Power, for example (1664), referred to "this particular Engine we call the Body" and to "the Wheelwork and Internal Contrivance of such Anatomical Engines," and claimed of "Insectile Automata" that, "in these prety Engines (by an incomparable stenography of Providence) are lodged all the perfections of the largest Animals."[169] Joseph Glanvill suggested that man's phantasies, "like the Index of a Clock, are moved but by the inward Springs and Wheels of the corporal Machine."[170]

Distinguishing between vegetable and animal nature, Matthew Hale compared the one with "a curious Engin . . . [of] some simple and single motions, like a Watch that gives the hour of the day" and the other with "a Watch that besides the hour of the day gives the day of the month, the age of the moon, the place of the Sun in the Zodiack, and other curious Motions wrought by multiplication of Wheels."[171] And Thomas Burnet spoke of "the Stomach and the Heart as the two Master-Springs in the Mechanism of the Animal" and of "the Heart, which is the chief Spring of the whole Machine."[172] Boyle called the animal or human body variously an "engine," a "living engine," an "admirable" or "admirably framed engine," "a very compounded engine," or a "living automaton," and exclaimed that "I have seen elephants, and admired them less than the structure of a dissected mole . . . [for] I must confess my wonder dwells not so much on nature's clocks . . . as on her watches."[173]

Such casual uses of clock metaphors in connection with animal life did not seem to offend anybody. Controversy, however, resulted from the logical extension of such practice, which was to declare, formally, as Descartes had done, that the entire animal was an automaton, an automaton in the strict sense, without mind and soul. And once this hypothesis of animal automatism was accepted, it was difficult not to take the last step and to classify humans, too, along with all the other animals, as machines. On the Continent, the doctrine of animal automatism was widely accepted and only when La Mettrie extended it to include man, did appreciable dissent arise. In Britain, by contrast, even early suggestions of animal automatism were rejected from the start.

Characteristic is the case of Henry More, clergyman, lifelong fellow of Christ College at Cambridge, Neo-Platonist, and acquaintance of Newton; he is credited with having done more than anyone else for the introduction of Cartesianism to Britain. In the first stage of his enthusiasm, he conducted a substantial correspondence with Descartes. Besides expressions of unrestrained adulation (he later turned into a violent prosecutor of Cartesianism, whose originator he called "a daring monster"),[174] he firmly took exception to the notion that animals should be mere machines: "My spirit . . . turns . . . with abhorrence . . . from that deadly and murderous sentiment . . . whereby you . . . withhold life and sense from all animals, for you would never concede that they really live."[175] He added a number of sentimental anecdotes about birds and dogs to show that animals have consciousness, understanding, and reason. This early reaction anticipated the subsequently conventional British position on this question.[176] A modern investigator of this question has concluded that "with the exception of Sir Kenelm Digby, who, as early as 1645, struck a good Cartesian stand on the question, and Anthony Le Grand, the exponent of Cartesianism in England, English writers are aligned against strict animal automatism."[177]

If the doctrine of animal automatism was rejected so unanimously, the farther reaching proposition of the man-machine was never taken seriously. It may be instructive to survey briefly Boyle's views on the matter; at once an enthusiastic mechanist and a devout Christian, perhaps he spoke for a majority. To Boyle, the bodies of animals and humans were unquestionably mechanisms but mechanisms of a higher order than clockwork: "No watch or clock in the world is in any way comparable, for exquisiteness of mechanism, to the body of an ass or a frog."[178] He repeated this at other occasions: "Man is not like a watch; the structure even of the rarest watch is incomparably inferior to that of the human body."[179]

The difference between the mechanisms of animal bodies and mechanical clocks was not one of degree but of kind; living organisms were "engines of God's own framing."[180] The main distinction was this: humans and, presumably to a lesser degree, animals had a capacity of discernment, judgment, and free choice that distinguished them from automata and that was conventionally called "the soul." Boyle realized that the word *soul,* apart from its theological meanings, disguised phenomena that were insufficiently understood.[181]

Soul, then, was a label for the unexplained interactions between body and brain, but it had yet another meaning. Traditionally the concept, soul, had served as symbol and criterion for that complex dynamic condition of the animal mechanism known as *life.* Whoever attempted to describe the animated body as a machine had to come to terms with that mysterious phenomenon. Boyle, it seems, viewed the living organism as a mechanism of vast complexity, and as a system of interacting processes that as yet were largely unexplored but that might yield to the methods of future science.[182] The phenomenon of life, then, was the result of a complex collaboration of physiological agents, some of them elusive and yet to be identified but nevertheless in principle accessible to empirical investigation. Life was not, he implied, the manifestation of some ineffable spiritual entity such as the soul.

The extraordinary fondness that Boyle displayed throughout his writings for the clock and the clock metaphor makes the firmness with which he rejected applications of the clock image to the living bodies of humans and animals all the more remarkable. To some extent he was motivated, no doubt, by traditional theological loyalties. He must have sensed that no clear criteria existed for distinguishing between animals and humans, and he also must have recognized that Descartes's attempt to avoid theological offenses by maintaining this distinction through use of the concept, soul, was doomed. Additionally, he must have feared that any association with the notion of the automaton might compromise the value of human free will. The subsequent career of the doctrine of animal and human automatism in England has been competently dealt with in a specialized literature,[183] and since it does not add significantly to our understanding of the history of the clock metaphor, it will not be pursued further here.

Determinism versus Free Will

The mechanical philosophy was concerned with metaphysics as much as with the physical world; its contributors consisted of theologians as well as of men trained in mathematics and the natural sciences. Since mechanics had no natural affinity with religion—if anything there was a suspicion to the contrary: Cartesianism was often charged with materialism and even atheism—mechanical philosophers had an understandable motive for emphasizing the harmony between the mechanist world view and Christian theology. What better way could there be to do this than to prove the existence of God by mechanical reasoning?

There was nothing new, to be sure, in the argument from design; its general form had been inherited from antiquity, and its specific association with the clock metaphor went back to the early fifteenth century (please see pp. 30, 48, and 57). In the seventeenth century, with the rise of the mechanical philosophy, the argument from design began to be invoked with great frequency, both in full formulation and in abbreviation, on the Continent and in Britain. Its subsequent career on the Continent was unremarkable, but in Britain it became the vehicle of an extraordinary discussion about the attributes of God and the merits of determinism and free will.

In its standard form, the argument from design was patterned after the simple syllogism:

> Clocks are made by clockmakers and not by chance.
> The world is like a vast clockwork.
> Therefore it, too, must have its specific maker.[184]

In this simple form, the argument was used both in Britain and on the Continent. While more persuasive, perhaps, through its appeal to sentiment than to logic, this argument was offensive to no one.

A cause for controversy presented itself, however, when English writers (an early example was John Robinson—see chap. 2, pp. 48–49), reading the implications of the argument more closely, discovered the dismaying dilemma that God's omniscience and omnipotence were mutually exclusive. If God was all wise, he could not be all powerful, for, having produced a perfect creation that is capable of running flawlessly ever after, he had made himself superfluous, ending up in the role of the passive spectator. Conversely, if God was all powerful, it was essential that he continue to govern actively the world after its creation; a world that needed continued management, however, could not be perfect, and its creator could not be all wise. The ensuing debate, the opposing camps of which are now labeled "intellectualists" and "voluntarists," was in essence a variant of another fundamental debate of the age, namely, that between determinism and free will, or authority and liberty.

The design argument began to appear frequently in British mechanist writings in the middle of the seventeenth century, in direct consequence, it seems, of the new popularity of Descartes. The first formulations of the argument unhesitatingly took the intellectualist point of view, which saw the creation of the world in direct analogy to the making of a clock: if a man-made clock had all its future actions programmed into it by its maker and would run—if perfectly made—forever without need of repair, the clockwork of the universe, made by a divine creator of absolute and unlimited wisdom, could not be capable of less. Early testimony to this view comes from two Englishmen who were instrumental in introducing Descartes to the British public, Sir Kenelm Digby and Henry More.[185]

The same position is underlying an argument offered by Henry Power (1664) to refute predictions of the impending end of the world (based on biblical information on the age of the earth—according to which Bishop James Ussher had dated the Creation at 4004 B.C.—and the astronomical fact that the precession of the earth's axis needed over

20,000 years to complete a full cycle) on the grounds that "he that made this great Automaton of the World, will not destroy it, till the slowest Motion therein has made one Revolution."[186]

Boyle, the chief spokesman for the mechanical philosophy, invoked the design argument innumerable times. In accord with his inclination to hold middle-of-the-road positions, the changes in his formulations of the argument subtly reflect the shifting consensus of the British intellectual community. At first, in the early 1650s, Boyle took a position that was unambiguously intellectualist and expressly antivoluntarist, a position that he himself later, in the hindsight of subsequent debates, might have judged naïve. He said, in brief, that the world was created so perfectly as to cause the illusion of an "intelligent being, watchful over the public good of it," constantly directing and maintaining it; this illusion, he added, was comparable to believing that the multiple performances of the monumental clock of Strasbourg Cathedral were caused by a deliberate conspiracy of its various parts.[187]

Often in his earlier writings, Boyle had expressed his belief in determinism by comparing the world to clockwork; for example, he referred in 1665 to the mechanical philosophy as "that hypothesis, supposing the whole universe (the soul of man excepted) to be but a great Automaton . . . , [where] all the phenomena result from those few principles, single or combined, (as the several tunes or chimes, that are rung on five bells,)."[188]

The issue was raised again at length in *A Free Inquiry into the Vulgarly Received Notion of Nature* (published in 1686 but written around 1665 or 1666). Now Boyle's determinism was not only circumscribed by the formula, "the soul of man excepted," but also generally showed signs of softening toward voluntarism. God's arbitrary intercession in the functioning of the world was no longer actually ruled out. But while Boyle conceded God the right of such intercession, he still declared it the mark of an artist's genius that his creation should be capable of functioning by itself, without the need of occasional help from some *deus ex machina*.[189]

He again rejected the view that sees God as constantly manipulating the affairs of the world by equating that view to the ignorance of simple-minded spectators of the Strasbourg clock who think its performances are individually produced by hidden operators.[190] A few pages later he once more asserted the supremacy of the divine intellect.[191]

The intellectualism expressed in *A Free Inquiry* was tempered, however, by explicit allowances for God's omnipotence and the free exercise of Divine Will. It was by his free choice if God normally did not intercede in the harmonious running of the clockwork universe, which, of course, he could do if he wished. Man's spiritual life, moreover, was in principle exempt from the determinism of the physical world.[192]

The insistence on human free will is connected with an attempt to balance God's omniscience with his omnipotence, a formula that tries to strike a reasonable compromise between the intellectualist and voluntarist positions.[193] While Boyle was making a gradual and cautious transition from an intellectualist to a voluntarist position, others began to argue the voluntarist case aggressively. Henry Stubbe, in a literary feud with Joseph Glanvill, observed in 1670,

> that if *God Almighty* be regulated by the rules of Geometry, and *mechanical motion* in the management of this *world,* and that the *fabrick* of things is *necessarily* established upon *those Hypotheses,* I cannot any way comprehend how *God* can do any *miracles:* how the *Sun* should *stand still* at the command of Joshuah.[194]

What was expressed here only in a casual remark became the central theme of Ralph Cudworth's *True Intellectual System of the Universe* (1678), which was expressly designed as "a Discourse concerning Liberty and Necessity, or to speak out more plainly, Against the Fatall Necessity of all Actions and Events."[195] In this campaign against determinism, Cudworth identified the true enemy in the philosophy of Descartes, a philosophy that he knew well and from which he had benefitted a great deal. Charging the Cartesian deists with having "an Undiscerned Tang of the Mechanick Atheism hanging about them, in . . . their so confident rejecting of all *Final* and *Intending Causality* in Nature,"[196] he rejected the notion that all earthly events are exclusively predetermined by an initial set of efficient causes because that view was in conflict with his belief in an omnipotent God whose freedom of action was unlimited and who directed the fortunes of the world according to his "providence" as a set of final causes.[197]

Combining the idealism of Cambridge Platonism and Protestant orthodoxy with the language of mechanical philosophy, Cudworth's *True Intellectual System* won a sympathetic reception and exerted considerable influence. With this book, the voluntarist point of view was unassailably established in Britain.

When a lonely dissenter, Thomas Burnet, tried in 1684 to defend the intellectualist-determinist side of this debate ("We think him a better Artist that makes a Clock that strikes regularly at every hour from the Springs and Wheels which he puts in the work, than he that hath so made his Clock that he must put his finger to it every hour to make it strike"),[198] he was met by solid and vociferous opposition.[199]

Boyle adjusted himself to this new reality by adopting a voluntarist position that was tempered by certain elements of his earlier intellectualism. When he summarized his view of the design argument and the issues raised by it in his last major work, *The Christian Virtuoso,* he did so without reference to clockwork.[200]

Although, as the quotations from Cudworth and Burnet show, the clockwork image was generally regarded as an attribute of intellectualism, voluntarists, too, continued to make use of the old image for some time. George Cheyne, a minor Newtonian philosopher, compared God's continuing active participation in the affairs of the world with the periodic winding up of clocks. The universe, he maintained,

> requires an *extrinsick Principle* for its subsisting in its present condition. If one should see a Piece of *Clock-work,* pointing out the Divisions of Time exactly and regularly, he might have some Difficulties about the manner of its Production; but if he should see or learn, that it required some *Foreign Assistance* to keep it going, that its Motion depended upon some Principle without itself, that it required winding up of the Springs or Weights . . .[201]

Several further references to clock imagery in Cheyne's book bespeak a fondness of that machine; however, this did not affect his scale of values. While he insisted on the unlimited freedom and power of God, he also believed in the reality of human free will. And these things had nothing in common with the clock:

> That *Freedom* and *Liberty* of choosing and refusing which we find in ourselves, is altogether inconsistent with *Mechanism.* Some Men indeed deny that we have any Free-will at all; but these need only examine their own Consciences to be convinc'd of their mistake.[202]

William Derham, a theologian and naturalist who had written a workmanlike technical manual on clocks,[203] combined the clock metaphor with an insistence of God's omnipotence in the bland assertion that "this [the order and regularity of the world] is a manifest sign of a wise and kind, as well as omnipotent CREATOR and ORDERER of the World's Affairs, as that of a Clock, or other Machine is of Man."[204]

Clock imagery also appeared in George Berkeley's early (1710) *Treatise concerning the Principles of Human Knowledge.* In that work, God's omnipotence was assumed from the start. The question was only: why did God create the phenomena of the visible world through the complex processes of nature if he could, as some argued, achieve the same results by a direct "fiat or act of His will," that is, by miracles?[205]

His answer was that God could indeed "produce a miracle, cause all the motions on the dial-plate of a watch, though nobody had ever made the movements and put them in it." But this God does that only rarely. Normally,

> God seems to choose the convincing our reason of His attributes by the works of nature, which discover so much harmony and contrivance in their make, and are such plain indications of wisdom and beneficence in their Author, rather than to astonish us into a belief of His Being by anomalous and surprising events.[206]

The purpose of nature, this comes close to saying, is to testify to the attributes of God, and the purpose of clockwork is to illustrate the relationship between the Creation and its Maker.

Sir Isaac Newton never explicitly used the clock metaphor. Deeply religious, if reticent about his beliefs (some of which were lacking in orthodoxy), he referred to the design argument frequently. Although some formulations paid pointed tribute to God's wisdom, Newton's commitment to voluntarism was unequivocal. In the General Scholium of the *Principia,* for example, he wrote, "This most beautiful system of the sun, planets, and comets, could only proceed from the counsel and dominion of an intelligent and powerful being."[207] And in a letter to Bentley (10 December 1692), "Ye motions wch ye Planets now have could not spring from any naturall cause alone but were imprest by an intelligent agent."[208] Similarly, in *Opticks* (1704), "Blind Fate could never make all the Planets move one and the same way in Orbs concentrick. . . . Such a wonderful uniformity in the Planetary System must be allowed the Effect of Choice."[209]

Newton's voluntarism was not a position that he received uncritically from his intellectual environment but a strong personal conviction that he defended with energy. Unlimited power, Newton held, was God's chief characteristic:

> This Being governs all things, not as the soul of the world, but as Lord over all; and on account of his dominion he is wont to be called *Lord God* παντοκράτωρ, or Universal Ruler; . . . a being, however perfect, without dominion, cannot be said to be Lord God. . . . A God without dominion, providence, and final causes, is nothing else but Fate and Nature.[210]

The proof of such unlimited power was God's unbounded freedom to exercise his will. Newton therefore characterized the creation as "the Effect of Choice"; the world consisted of "Creatures subordinate to him [God], and subservient to his Will."[211]

Newton's defense of God's omnipotence, while inspired by theological considerations, was reinforced by his judgment as a scientist. He was persuaded that the cosmos could not run forever without God's periodic intervention; he speculated a great deal about the specific flaws that would make such intervention necessary and the nature of the required intervention.[212] In *Opticks,* he pointed to "some inconsiderable Irregularities" in the planets' orbital travel "which may have risen from the mutual Actions of Comets and Planets upon one another, and which will be apt to increase till this System wants a Reformation." He was also concerned that the total "quantity of motion" in the world was gradually diminishing: "Motion is much more apt to be lost than got, and is always upon the decay."[213] To a friend he confided:

> that a continual miracle is needed to prevent the Sun and the fixed stars from rushing together through gravity: that the great eccentricity in Comets in directions both different from and contrary to the planets indicates a divine hand: and implies that the Comets are destined for a use other than that of the planets. The Satellites of Jupiter and Saturn can take the places of the Earth, Venus, Mars if they are destroyed, and be held in reserve for a new Creation.[214]

On another occasion, he declared that "the fixed starrs not coming together is a constant miracle."[215]

Newton's universe was clearly not an idealized clockwork that ran flawlessly forever without rewinding or repair but a constantly changing dynamic system needing constant attention and periodic adjustment from God. Newton may have thought of practical clocks with their inaccuracies and breakdowns, but he never used the clock metaphor and limited himself to such expressions as "the frame of the universe" or "of nature."[216] The slogan, "Newton's clockwork universe," which modern writers are fond of repeating, then, is not only unsupported by the sources but is also inconsistent with Newton's views.[217]

Newton's voluntarism, that is, his conviction of the supremacy of God's power, will, and providence was not a personal eccentricity held by him alone but a belief that had developed in England over some time and that was shared by the leading thinkers there. Newton only lent this view the support of his towering prestige. At the same time, Continental thinkers were oriented along intellectualist lines. The two opposing camps were thrown into direct conflict by a curious chain of events.

In 1714 Georg Ludwig, elector of Hanover, left his German court to become King George I of England. Gottfried Wilhelm Leibniz, the elector's librarian and historian who was left behind in Hanover, kept up a correspondence with his patron's daughter-in-law, Caroline, now princess of Wales and an old friend and disciple; Leibniz probably hoped for an invitation to rejoin the court in London. This wish naturally motivated Leibniz to point out to the princess the weaknesses of the philosophic system then predominant in Britain, namely, that of his old rival and antagonist, Sir Isaac Newton. In November 1715 Leibniz sent the princess a letter listing his chief objections. The princess, a well-educated, intelligent woman with a genuine interest in philosophy, became curious to hear the other side. She showed the letter to Dr.

Samuel Clarke (1675–1729), a promising young theologian at court and a loyal disciple of Newton.

Clarke, in consultation with his mentor, composed a reply to Leibniz, thus setting off a famous debate that ultimately ran to five such exchanges and ended with Leibniz's death. Viewed not only as an admirable summary of the two chief philosophies of the day but also as a spectacle of duelling giants, it was published at once and reissued many times.[218]

Newton's spokesman, Samuel Clarke, although virtually forgotten today, had a considerable reputation at the time. Clarke had embraced Newton's system early and performed some important services in its cause. He turned Rohault's *Physics,* the standard text of Cartesian physical doctrine, into a statement of Newtonianism by publishing a new Latin translation of it (1697) that featured extensive notes explaining the Newtonian position on every issue. He also prepared a Latin translation of Newton's *Opticks* (1706) which was enthusiastically authorized by the master. In the debate about freedom versus determinism, he had already established himself as a forceful champion of personal liberty and free will.[219] And he energetically rejected the clock metaphor as early as 1708:

> If the Mind of Man were nothing but a certain System of Matter; and Thinking nothing but a certain Mode of Motion in that System: It would follow, that, since every Determination of Motion depends necessarily upon the Impulse that causes it, therefore every thought in a Man's Mind must likewise be necessary, and depending wholely upon external Causes; And there could be no such thing in Us, as Liberty, or a Power of Self-Determination. Now what Ends and Purposes of Religion mere Clocks and Watches are capable of serving, needs no long and nice consideration.[220]

It would be neither fitting nor necessary to survey the entire debate here; it will be sufficient to sketch only the exchanges that pertain to the clock metaphor.[221] Concerning Newton's direct role in the controversy, there is evidence of continuous consultation and of some active if anonymous participation.[222] Nevertheless, Clarke's role was not that of Newton's unthinking servant but rather that of an independent thinker who happened to share and admire Newton's views.

As the debate was opened by Leibniz, it was also he who introduced the clockwork metaphor. He did so in a strongly worded challenge of Newton's views on God's way of governing the universe:

> Sir Isaac Newton, and his followers, have also a very odd opinion concerning the work of God. According to their doctrine, God Almighty wants to wind up his watch from time to time: otherwise it would cease to move. He had not, it seems, sufficient foresight to make it a perpetual motion. Nay, the

> machine of God's making, is so imperfect, according to these gentlemen; that he is obliged to clean it now and then by an extraordinary concourse, and even to mend it, as a clockmaker mends his work; who must consequently be so much the more unskilful a workman, as he is oftener obliged to mend his work and to set it right. According to my opinion, the same force and vigour remains always in the world, and only passes from one part of matter to another, agreeably to the laws of nature, and the beautiful pre-established order. And I hold, that when God works miracles, he does not do it in order to supply the wants of nature, but those of grace. Whoever thinks otherwise, must needs have a very mean notion of the wisdom and power of God.[223]

What Leibniz offered in place of Newton's alleged errors was the intellectualist image of the clockwork universe with God as its infallible clockmaker, an image that Leibniz had made the basis of his system of the preestablished harmony. Clarke answered with the familiar voluntarist argument that God's unlimited ability to exercise his power and his sovereign will were his most important characteristics and that to deny this would be to approach atheism. If someone claimed that a kingdom was so well governed that the king had nothing to do, that person should rightly be suspected of wanting to depose the king. More remarkable, however, is the main part of Clarke's reply:

> The reason why, among men, an artificer is justly esteemed so much the more skilful, as the machine of his composing will continue longer to move regularly without any farther interposition of the workman; is because the skill of all human artificers consists only in composing, adjusting, or putting together certain movements, the principles of whose motion are altogether independent upon the artificer: such as are weights and springs, and the like; whose forces are not made, but only adjusted, by the workman. But with regard to God, the case is quite different; because he not only composes or puts things together, but is himself the author and continual preserver of their original forces or moving powers: and consequently 'tis not a diminution, but the true glory of his workmanship, that nothing is done without his continual government and inspection. The notion of the world's being a great machine, going on without the interposition of God, as a clock continues to go without the assistance of a clockmaker; is the notion of materialism and fate, and tends, (under pretence of making God a *supra-mundane intelligence*), to exclude providence and God's government in reality out of the world.[224]

Clarke's words "with regard to God, the case is quite different" broke the spell that the clock metaphor had held over Europe for many generations. To identify the clock image as an accessory of determinism and, by implication, an enemy of liberty was as accurate as it was effective before an English audience. More deflating to the power of the metaphor, however, was to state boldly on logical grounds its irrelevance to the entire issue of God and the Creation in the first place.

Clarke's attack proved effective. Leibniz never replied to the charge of the clock metaphor's irrelevance, even when Clarke repeated it for good measure in his Second Reply: "The case of a human workman making a machine, is quite another thing" [than "the workmanship of the universe"].[225] Instead, Leibniz belabored the conventional intellectualist argument that wisdom, not power, was God's supreme characteristic. In subsequent exchanges, the question of priority of God's power or wisdom was continued with increasing wordiness. After the second exchange, however, the clock, either expressly or by implication, was not mentioned again in the debate.

4

The Clockwork State

Since Plato's *Laws,* Aristotle's *Politics,* and Livy's legend of Menenius Agrippa, Europeans have been accustomed to comparing the state with the human body.[1] In the late Renaissance, while this metaphor continued to flourish undiminished, others appeared by its side, for example, the images of the "ship of state," of the planetary system ruled by the *primum mobile* or, after Copernicus, by a monarchical sun, and, by the late sixteenth century, the metaphor of the clock.[2]

A few early authors offered the clock as an illustration of the king's function in the realm: being visible to everyone, he gave precept and direction to all his subjects. Dealing only with the *external* appearance of the clock and based on a simplistic and somewhat sentimental notion of government, this picture did not focus on the essential features of either clock or state; it had little influence.

More compelling, however, was another form of the clock metaphor which concentrated on the *internal* workings of clockwork and state. Examples of this use of the metaphor began to appear at a time when mechanical clocks, produced in greatly increased quantities due to the consolidation of the clockmakers' craft, became accessible to a wider public, when the clock metaphor in its manifold and diverse applications became ubiquitous in the general literature, and when the absolute monarchy was generally considered the best form of government.

In 1572, Detlev Langebek (1510–75), a North German author of a book on statecraft, described human society as a clockwork with various large and small wheels.[3] Justus Lipsius (1547–1606), one of the earliest theorists of absolutism, compared the contrast between the inscrutability of the inner workings of government and the conspicuous visibility of the ruler's public life with that between the internal mysteries of complex clockwork and the simplicity of the action of the clock hand.[4] In 1634 Henri Duc de Rohan (1579–1638), a notable French soldier-statesman, called the Spanish state "a huge machine composed of divers parts . . . encumbered by its own weight that moves by its secret spring." Although he did not say what machine he had in mind, his choice of words suggests clockwork.[5]

Antonio Ponce de Santa Cruz (d. 1650), physician to the Spanish king, invoked the similarity of clock and state to clarify the difference between the Aristotelian concepts of "conjoined and separate instruments.":

> He explains the Nature of a separate Instrument by an Example political. For a King or some Potentate operates in far distant places as if he were present, by power derived from himself, which he confers upon the Laws and Judges. . . . He brings afterward another example whereby he would prove his opinion, viz., Of Watches, Clocks, and Engines wherein many Wheels are orderly moved in the absence of the Workman, yet by a vertue imprinted upon them by the first direction of the Artist.[6]

The particular feature of the monarchical state, then, that reminded Ponce de Santa Cruz of clocks and automata was its command structure.

Two authors of books on emblemata described the kinship between monarchical government and the clock even more graphically. Florentius Schoonhovius (1618) explained the motto "sat cito si sat bene" (fast enough if good enough) by pointing to an accompanying illustration of a mechanical clock:

> As the diverse swinging in opposite direction of the circles holds together the passionate motion of the wheels, so does the farsighted caution of old age rein and steer the hot daring of youth. Therefore the prince should choose old men to advise him; the younger ones he should assign to the execution of the judgement of the wise elders.[7]

The motto "uni reddatur" (it [power] shall be given to only one) prompted Diego de Saavedra Fajardo (1649) to construct the following comparison between clockwork and monarchy. After describing the harmonious cooperation between the hidden gears and the widely visible hands of a clock, he stated:

> Such harmony should prevail between a prince and his councillors. . . . Monarchy is distinguished from other forms of government in that only one commands, all others however obey. . . . Therefore in the clockwork of government the prince should be not only a hand but also the escapement that tells all other wheels the time to move.[8]

Such imagery expressed accurately the aspirations of the rising political doctrine of absolutism: government should work with the harmony of a smoothly running clockwork: The power of the sovereign was to be unlimited and absolute, and government was to be centralized with the king as the source of all power and as the author of all initiative.

Sentiments like these would have been congenial to a contemporary political philosopher in Britain, Thomas Hobbes (1588–1679), who put the automaton analogy into the center of an ambitious philosophical system. His thinking had been deeply affected by the mechanical ideas of contemporary natural philosophers, especially Descartes with whom he was linked by personal animosity as well as by a deep-seated if unacknowledged intellectual affinity. Hobbes had a reputation as a most radical mechanist himself. He was one of the few in England who accepted Descartes's theory of animal automatism and, believing in determinism and rejecting free will, he easily extended that theory to include man.

Hobbes's main works on political theory, *De Cive* and *Leviathan,* employed the clock metaphor in prominent roles. In the earlier work, *De Cive* (1642), the clock image served to introduce the elementary rules of mechanical philosophy into political analysis. Hobbes restated such familiar mechanistic rules as unknown phenomena should be reduced to the quantities of "matter, figure, and motion" and a complex problem should be analyzed by taking it apart into its smallest elements in the manner of a clockmaker.[9]

More novel was Hobbes's use of the clock image in the *Leviathan* (1651). He was familiar with both the popularity and the efficacy of the traditional "body politic" analogy for the state and incorporated this analogy into the mechanistic program by simply expanding it to include the clock, that is, by equating both the living body and the state with the clockwork-driven automaton.[10]

Hobbes introduced this state-automaton analogy explicitly only at the beginning of the *Leviathan,* but in method and spirit it was present throughout the book. He analyzed the components, forces, and actions of the state with the detachment of a mechanician. When he listed at length the physiological counterparts for the various components of the social organism, the analogous parts of a clockwork were implied and could easily be guessed.[11]

The purpose of Hobbes's analysis was to synthesize an artificial state of great structural stability, a state, that is, equipped above all to maintain its civic peace. His result was a centralistic, hierarchical form of government, where the actions of the lower members were controlled through rigid administrative linkages by the sovereign above. It was an authoritarian state in which individual citizens had no choice and no vote. Hobbes had presumably been driven to this authoritarian view by experiences of recent history, such as the British Civil War, the chaos in France after the assassination of Henri IV, and the Thirty Years' War in

Germany. But might not his way of envisioning an ideal organization also have been reinforced by the paradigm of the clock?

As far as the use of the clock metaphor is concerned, little about Hobbes's influence can be said with certainty. After him, although not necessarily due to his efforts, the old metaphor of the body politic and the new one of the clockwork state seem to have been well reconciled. Subsequent authors, especially on the Continent, often used the two next to each other and occasionally fused them into outright "mixed" metaphors. Concerning Hobbes's more general impact, especially on contemporary Britain, it is not obvious that his precedent helped to advance the use of mechanical reasoning and imagery in political theory. A century later, when Continental political theory had indeed become thoroughly mechanical, there were no acknowledgments of any debts to Hobbes.

By the end of the seventeenth century, political thought in Britain was developing in a somewhat different direction than on the Continent, and the difference was reflected in the fortunes of the clock metaphor. In English political literature, the use of mechanical imagery declined. Explicit clock metaphors became rare and tended to have anti-mechanical connotations. For example, William Penn, in his *Frame of Government of Pennsylvania* (1682), had observed:

> Governments, like clocks, go from the motion men give them; and as governments are made and moved by men, so by them they are ruined too. Wherefore, governments rather depend upon men than men upon governments. Let men be good and the government cannot be bad; if it be ill, they will cure it. But if men be bad, let the government be never so good they will endeavor to warp and spoil it by their turn.[12]

This was not praise of machinelike government. By reminding us that even the most carefully devised governments are composed only of humans, Penn pointed out that what they had in common with clockwork was frailty and imperfection.

Other less specific mechanical imagery can occasionally be found in British political writings over the following century. George Savile (1688), for example, called a prince who was manipulated by his advisors "rather an Engine than a living Creature" and referred to various intrigues as "Engines of Dissention" and "inward Springs and Wheels whereby the Engine moved."[13] John Locke (1691) spoke of "money, in its circulation, driving the several wheels of trade."[14] And Walter Moyle (1699) wrote, "These politic orders were the great springs and wheels upon which this mighty fabric turned: but as all natural bodies are born with seeds of dissolution in their own frame, so these great artificial bodies . . . ,"[15] a remark that typically mixed vague clockwork terminology with organic allusions. Late in the eighteenth century, Adam Smith was still using mechanical imagery in the same loose way. He spoke of "the wheels of the political machine" or of "the several wheels of the machine of government"[16] and observed that mercantilism's "two great engines for enriching the country, therefore, were restraints upon importation and encouragement of exportation" and that "monopoly is the great engine" of the American and East Indian trade.[17]

Such references to engines, machines, springs, and wheels were neither analogies nor metaphors, for they no longer suggested any specific mechanical device; rather they were shopworn figures of speech chosen precisely for their ambiguousness to refer to subjects for which the writer lacked affection. They were neither capable of nor meant for clearly defining, to say nothing of shaping, any ideas that were important to the writer.

On the Continent, however, the use of the clock metaphor to illustrate political ideas continued in the tradition established at the beginning of the seventeenth century. Until well into the eighteenth century, while not overly frequent, such metaphors were seriously descriptive and affirmative about their subjects.

Daniel Casper von Lohenstein (1635–83), a Silesian dramatist with strong political interests, was notably fond of the clock metaphor, especially as applied to matters of government.[18] In comparing the absolute ruler with the weight or escapement of a clock, Lohenstein made a shrewd observation. Although the two, one as the regulating device and the other as the source of energy, were technically not comparable, either one could be viewed as the central element of the whole machine upon which all other successful functioning depends. Lohenstein's contention was that in good government, just like in a clock, action flows unidirectionally outward from a central source to the various practical activities at the periphery.

A similar confidence in the concept of the clockwork state was expressed in the preface of Wilhelm von Schröder's (1640–88) book on the management of state finances (1686), in which he defended absolute government against the charge of arbitrariness with the argument that the critics "do not know how many large and small wheels make up a clockwork, all contributing to the motion of the hand: if one such wheel becomes defective, and someone immediately would smash it to pieces, would not its entire movement be stopped for a long time?"[19] In other words, in the cause-and-effect chain of clockwork as well as of government, every wheel is necessary, even the smallest.

Clockwork-state metaphors of this type can be found in moderate measure in the Continental literature of the seventeenth century. That they were not more frequent may have to do with the literary conventions of the political philosophers of that time. Their writing was generally oriented toward, and modeled after, the classical authors of scholasticism and antiquity. Their style of argument was factual and lawyerly. For evidence they presented historical examples; metaphors and analogies did not fit.

A change occurred in the early eighteenth century. A new movement emerged in Continental political thought that combined a commitment to the established principles of absolute monarchy with an enthusiasm for the teachings of recent, especially mechanical, philosophy. This new form of absolutism, sometimes referred to as "enlightened despotism," flourished for only a few decades and only on the Continent, but its marks upon the European political consciousness can still be felt today.

The enlightened version of absolutism may be characterized as a mechanical theory of government.[20] Its reasoning was directly influenced by the mechanistic natural philosophy of the preceding century, and its literature teemed with analogies and metaphors of machines, especially clockwork. Where, when, and by whom this union between absolutism and mechanism might have been brought about is by no means clear, but one of its first and most prominent representatives was doubtless King Frederick II, "the Great," of Prussia (1712–86).

Predecessors or preceptors from whom Frederick might have learned his mechanistic political philosophy are not apparent, but he was intimately acquainted with the leading enlightened, that is, rationalist and implicitly mechanist philosophers of his time. Men like Voltaire, D'Alembert, LaMettrie, Maupertuis, Euler, and Algarotti each spent long periods at his court. As soon as he had succeeded to the throne, he recalled Christian Wolff from the exile to which Frederick's father had banished the philosopher for his views of determinism. Upon LaMettrie's death, Frederick himself composed the official eulogy of the Berlin Academy of Science.[21]

With thinkers such as these, Frederick shared a generally determinist outlook and a mechanist vocabulary that could be applied to various conventional topics of philosophy. He would refer to the familiar automata legends and would recite the design argument postulating a Clockmaker God.[22] In discussing the choice between free will and determinism, he repeated Leibniz's old dictum that a God whose creation is subject to chance is no better than an incompetent watchmaker, and he invoked the automaton as the symbol of determinism, declaring himself ultimately in favor of determinism rather than free will.[23] He regarded the human body, not surprisingly for La Mettrie's protector, as a machine; with regard to the mind-body question, he wrote:

> Unfortunately, the mind seems to be only an accessory of the body. Whenever the organization of our machine is out of order, so is the mind; nor can matter suffer without mind participating in its suffering. This strict union, this intimate connection, seems to be a very strong proof in the opinion of Locke. That which thinks in us is certainly an effect or a result of the mechanism of our animated machine.[24]

It is one thing, however, to be conversant in mechanistic jargon and another systematically to introduce mechanistic concepts and imagery into political philosophy. It is not inconceivable that Frederick may have performed this leap. Whatever its origins,[25] he employed clockwork- (or machine-) state analogies throughout his adult life.

In his *Considerations of the Present State of the Body-Politic in Europe* (1736), the twenty-four-year-old Frederick introduced the clockmaker-statesman analogy with great directness:

> As an able mechanic is not satisfied with looking at the outside of a watch, but opens it, and examines its springs and wheels, so an able politician exerts himself to understand the permanent principles of courts, the engines of the politics of each prince, and the sources of future events. He leaves nothing to chance; his transcendent mind foresees the future, and from the chain of causes penetrates even the most distant ages. In a word, it is the part of prudence to know all things, in order that all things may be judged, and every precaution taken.[26]

Characteristic was his insistence that the successful ruler must exercise comprehensive control over all functions and actions of the state, present and future. Explicitly mechanistic attitudes, incidentally, were not irreconcilable with the view of the state as an organism (the above quotation comes from an essay on the body politic); at the same time Frederick declared that "the prince is for his people what the heart is for the body."[27] Frederick's ideal state, nevertheless, was deeply akin to clockwork, and "heart for the body" was equivalent to "escapement for the clock." In an essay about the role of laws in the state (1750), he wrote that

> a body of perfect laws should be the crowning achievement of the human spirit as regards the politics of government: one would observe there a unity of design and of rules so exact and so well proportioned that a state conducted by such laws would resemble a watch all of whose springs have been made for the same purpose; . . . everything would be anticipated, everything would be coordinated, and nothing would be subject to mishap: perfection, however, is not the province of mankind.[28]

Thirty years later Frederick still voiced the same beliefs, again simultaneously through the clock and body metaphors.[29]

The two analogies of clock and body expressed two different principles that were both essential for Frederick's conception of the state: on the one hand, the body analogy, which likened the prince to the head or heart, served to defend the central and supreme place of the prince in the hierarchy of the state. The clockwork analogy, on the other hand, expressed the ideal of a state where all problems, now and hereafter, could be solved by appropriate administrative mechanisms that were programmed in advance to take care of any eventuality.

Frederick's commitment to a mechanical philosophy of government was quite evident to his contemporaries. Early observers regarded Frederick's Prussia as the very embodiment of the clockwork state. Goethe in 1778, on his only visit to Berlin, reported home that "from the huge clockwork that unrolls before you, from the movement of the troups, you can deduce the hidden wheels, especially that big old [program] drum, signed F[redericus] R[ex], with its thousand pins which generates these tunes, one after another."[30]

A generation later, a Hanoverian official who strongly opposed "the idea that a state is a machine constructed by the highest power" regarded Frederick as that idea's author: "In the army many things of necessity have to be arranged in a machine-like manner, and from the army Frederick received his notions of state. . . . When he . . . took the route of all despots, transferring the military mechanism to civilian administration, he allowed his . . . civil servants as little legitimate room for movement as possible and restricted the freedom of his subjects . . . to the utmost."[31]

Among Frederick's immediate contemporaries, the conception of the machine state had a considerable following. Partisans of this ideology appeared in both Germany and France. An early French expression of the mechanist attitude toward government was given in Condillac's *Traité des Systèmes* of 1749: "A nation is an artificial body; it is up to the magistrate who watches over its preservation, to maintain the harmony and strength in all limbs. He is the machinist who must repair the springs and rewind the whole machine as often as circumstances require."[32] The successful ruler, this was to say, must possess technical expertise in all aspects of statecraft at least at such a level as is expected from a watchmaker to whom one entrusts a complicated mechanism.

Mechanical imagery played a more central role in the writing of Jean Jacques Rosseau (1712–78), himself the son of a clockmaker. As the author of one of the most influential books on political philosophy of the century, he saw himself as an engineer of political machinery. At the beginning of an early version (1754–55?) of his *Social Contract,* he declared: "Therefore it is not here a question of the administration of this body, but of its constitution. I make it live, not act. I describe its springs and its parts, I arrange them in their place. I put the machine in a condition to run; others who are wiser will control the movements."[33]

In less personal terms and therefore without modest qualifiers, this was repeated in the final work (1762): "But if it is true that a great prince is a rare man, what then will a great lawgiver be? The first has only to follow the model which the other has to create. The latter is the mechanician who invents the machine, the former is merely the workman who winds it up to make it run."[34]

Rousseau employed such mechanical imagery with great frequency, both in casual remarks and extensive analogies. He compared, for example, the vitality of a form of government to the degree to which a clock spring is wound up.[35] When discussing ways of solving a constitutional crisis, he offered this illustration: "The machine must contain all the springs that should make it run: when it stops, the workman has to be called to wind it up again."[36] And he also saw similarities between governments and machines in their normal operation: both suffered from the phenomenon of friction.[37]

Rousseau's mechanical imagery reveals an ambivalent attitude toward the question of freedom. In principle he upheld and defended the idea of freedom, but in practice this led him into some tortuous reasoning. The social compact implied, he thought, that "whoever should refuse to obey the general will shall be compelled to it by the whole body: this in fact forces him to be free; for this is the condition which, by giving each citizen to his country, guarantees his absolute personal independence, a condition which gives motion and effect to the political machine."[38] Whatever Rousseau said about freedom, his machinery of state was centralistic; authority was to rule without dispute, and problems were to be solved by intervention of the government or, better yet, to be prevented by accurately planning ahead.[39]

In the manner of many absolutist political thinkers, Rousseau's mechanistic imagery appeared side by side with the ancient body-politic metaphor, invoked in support of central authority. In an article on "Political Economy" contributed to Diderot's *Encyclopédie,* Rousseau constructed a detailed analogy between the political organism and the human body. The sovereign was the head; the sense organs the judges and government officials; trade, industry, and agriculture were mouth and stomach; money the blood; and so forth. "The citizens form the body and the limbs that make the machine move, live, and work, and which cannot be injured in any part without the pain being transmitted to the brain at once, as long as the animal is healthy."[40]

Given this preference for centralized authority, it is only consistent that Rousseau's favorite form of government should be the monarchy.

Its functioning was described in terms of mechanical imagery.[41] And to illustrate the power of the king in an absolute state, and the absolute dependence of all subjects upon their sovereign, Rousseau pointed to the wizardry of Archimedes, history's most famous mechanician.

In Germany, the influence of Frederick's mechanistic conceptions of government was not confined to his own state. Even non-Prussian political authors were infected by his imagery as, for example, F. K. von Moser (1757), who was concerned with "keeping the clock of government going at the right pace and motion,"[42] and L. B. M. Schmidt, whose *Theory of Political Economy* (1760) derived its very structure from the machine-state analogy.[43] Frederick's most faithful disciples, however, were Prussians themselves. Johann Heinrich Gottlieb von Justi (1717–71), who concluded his service to the Prussian state—and his life—in a Prussian jail, was one of the chief representatives of Kameralism, the peculiarly German approach to economic and industrial planning by state authority.

Justi's writings are saturated with references to the "driving springs" ("Triebfedern," namely, virtue, honor, or love of equality, as the case may require) "by which the clockwork of the state is powered"[44] and with more elaborate state-machine analogies where the machine in question is readily identified as clockwork.[45]

There were two characteristics that to Justi specifically linked the concepts of machine and state: order and centralized control.[46] This directly leads to the responsibilities of the ruler: "His chief concern must be to be on constant guard to maintain this order. This is his noblest, indeed, his only duty. He is the governor of the machine of the body of state."[47] Elsewhere Justi expressed this thought even more pointedly: "A well-constituted state must perfectly resemble a machine where all wheels and gears fit each other with the utmost precision; and the ruler must be the engineer, the first driving spring or the soul . . . that sets everything in motion."[48]

The same opinion was held by Justi's contemporary and compatriot, Johann Friedrich von Pfeiffer (1718–87): "In the system of the monarchical state all driving springs, chains and wheels of the machine are united in one hand."[49] About recent experiments with constitutional forms of monarchy—obviously in England—he was sceptical: "They tinkered with the clock of government, and tried to make do with moderated or mixed monarchy." While the spectacular success of the British could not be denied, he doubted its permanence. Pfeiffer's conception of the state (states are "moral machines")[50] was a characteristic blend of the metaphors of body and machine.[51] Elaborating on the function of the ruler, he added: "What wisdom is needed to wind all springs of the machine of state properly and, if here and there they have lost their elasticity, either to restore them or to discover new ones."[52] Pfeiffer's summary of the absolute monarch's method of ruling is revealing:

> These are, roughly, the means by which the wise prince fits together the wheels and drive springs of the machine of government in such a way that all parts maintain a precise relationship to one another, and that none can neglect its duties without being noticed, all the more so since the prince can surround his office with accurate geometric drawings of all his states in order to quickly orient himself; with the help of general statistical charts that are to be submitted every six months he can at a glance, as it were, oversee the whole, make individual judgments as the case may require, and in the event of doubts or ambiguities request further explanation, or order special investigations at the site.[53]

This picture of the prince at his central command post supervising all the diverse functions of government by means of communication links that are direct but dependent on his initiative—a view similar to Descartes's automaton analogy for the brain-limb relationship in the human body—gives the clearest expression possible of the absolutist conception of the state.

A somewhat younger German representative of political mechanism was August Ludwig von Schlözer (1735–1809), professor at Göttingen and a notable Kameralist. His views of the state were similar to his predecessors' and just as explicit: "The state is first an invention: men made it for their own benefit, as they invent fire insurances, etc. The most instructive way of teaching the theory of the state is to treat the state as an artificial, exquisitely composed machine that is to run for a definite purpose."[54] Echoing Rousseau, he attempted a historical account for the evolution of this mechanism.[55] Consistent with the doctrines of absolutism and mercantilism, this machine was characterized by a centralized pattern of control.[56] Like other mechanists of state, Schlözer was deeply convinced of the need for constant vigilance in supervising the machine and for anticipation of all eventualities.[57]

In this sampling of eighteenth-century clockwork- (or machine-) state analogies, certain themes have emerged with conspicuous frequency and consistency: the need of the ever vigilant statesman-engineer who can predict as well as repair all potential troubles, the advantages of central coordination of all actions of the state machine in the hands of one director, the smooth running of clockwork as the ideal of order, and the rightness of a hierarchical distribution of the functions of state. The evidence presented here suggests that, roughly between 1740 and 1780, many and perhaps most German and French political thinkers were in the habit of visualizing the state, the government, and the body politic primarily in terms of clockwork.

Some subtle differences appear, however, between French and German uses of clockwork-state imagery. To German authors, the metaphors of clockwork and clockmaker tended to be more than mere illustrations of qualities desirable in government and statecraft. The mechanical imagery supplied its own imperatives; it insinuated values into politics that had no basis in the requirements of statecraft and that were recommended only by their efficacy in machinery.

Representative French writers, it seems, were free of this partisanship for machinery. That the state resembled a machine was accepted as an observed fact, a necessary evil one was not expected to like. On the contrary, if the state must be a machine, then at least it should be as simple as possible. Claude Adrien Helvétius likened "good" government to a "simple machine, whose springs would be easy to direct and would not require that great apparatus of wheels and counterweights that are so difficult to rewind."[58]

The Baron d'Holbach similarly postulated that "the great art of politics would be to act in such a way that in the complicated machine of society there would be no unnecessary, useless, or counteracting springs, but that all would conspire without wavering to the same goal."[59] When society, in a social compact, transferred its rights into the hands of a single sovereign, this was done in the interest of "simplifying a machine which had become too complicated through the opposing efforts of all its parts."[60] And to explain the unpleasantnesses involved in the life of a larger social community, he added: "The more complicated a machine, the easier it is to disrupt its course. The increased friction makes its running more delicate than that of a simpler machine."[61]

An even stronger dislike of the necessity to organize society as a machine was expressed by Diderot:

> I consider uncivilized human beings as a multitude of scattered and isolated springs. No doubt, if some of these springs should happen to collide, one or the other or both will break. To avoid such a mishap an individual of profound wisdom and high genius has assembled these springs and constructed from them a machine, and in this machine called society all springs are made to act and react on one another without ever stopping from fatigue. And more springs break in a day in a state of law and order [sous l'état de legislation] than would break in a year under the anarchy of nature. But what uproar! what waste! what enormous destruction of little springs when two, three, or four of these enormous machines bump into each other violently![62]

Diderot's distaste for the machinelike aspects of civilized society was theatrical enough, but he mentioned no alternatives; the existing state was inevitable. Nevertheless, his indictment anticipated future developments. The custom of Continental thinkers of representing society, government, and the state in terms of clock and machine analogies ran its course in the last quarter of the eighteenth century. The generation succeeding them not only rejected a state that resembled machinery; it also rejected quite expressly as being vicious the machine analogy itself.

5

The Authoritarian Conception of Order

Now that a tentative reconstruction, in an empirical, quasi-archaeological manner, of the career of the clock metaphor has been completed, it demands interpretation. What did those metaphors say? What do they mean? For whom did they speak? How is the history of the metaphor related to that of the mechanism itself? And what, finally, does the mechanical clock have to do with the original problem, the disregard for feedback mechanisms in Europe before 1700?

Metaphors and kindred devices of figurative expression are comparisons: some notion in need of illustration is compared with a different notion that must be familiar. The metaphor can be effective only if the two notions, the actual and the figurative, have some similarity despite their differences, some common characteristic, a *tertium comparationis*. A basic question is: within each individual clock metaphor, what is this common characteristic shared by the two notions under comparison? Which leads to a broader question: what do all the clock metaphors, despite their evident diversity, have in common?

Common characteristics do indeed emerge. The most important one is elementary. From the beginning, the image of the clock was linked with concepts that people held in high regard: the clock was called upon to illustrate the attributes of God; the harmony of the universe; the joys of paradise; temperance, the highest of the seven virtues; the truth of the new science; and the effectiveness of absolute monarchy. A machine that was not only popularly regarded as akin to such subjects but that was also invoked purposely to add to their praise clearly enjoyed the public's approval to an extraordinary degree. But such universal public approval, although an important fact, tells us little. We need to know more precisely in what way the clock was linked with the chief values of early modern European society.

A characteristic of the mechanical clock that seems to have aroused admiration from the start was the regularity of its running. Seeing a clock run, early viewers who, of course, had never known anything like it were enchanted by its harmony and orderliness, otherworldly qualities normally associated with things eternal and divine.

This connection between the clock and the Divine received added authority from having first been pointed out by a poet of the stature of Dante and by a widely read theologian like Suso. The perceived regularity of the clock's running (all-too-frequent breakdowns, it seems, were charitably overlooked) was also the basis for its identification with the recognized chief virtue of the age, temperance, an identification that became conventional both in literature and in art. And regularity became the basis for many other comparisons on less lofty levels, down to jibes for excessive punctuality.

An idealized clock, running for weeks and months with a constancy of pace that seemed unaffected by worldly troubles and that seemed to rival in regularity and dependability the eternal cycles of day and night and the motions of the stars, such a clock invited comparison with the world itself, and the comparison pointed to God as the divine clockmaker. From the late fourteenth to the late eighteenth centuries, this resemblance was observed innumerable times. It was made the basis of a formal proof of the existence of God, the "argument from design," which concluded that a divine creator must exist because the world has all the characteristics of a clockwork-like artifact. This analogy of clock and world, which was deeply and widely believed for centuries, came to play the role of a master metaphor from which most of the other clock metaphors were derived.

From the conviction that the world had the character of clockwork followed the belief that nature obeyed the laws of mechanics. This belief became one of the constituting elements of the new science and found expression in the often-repeated call that philosophers should approach the secrets of nature in the manner of the clockmaker who uncovers the causes of trouble by taking the malfunctioning mechanism apart.

The new science insisted that all of nature was subject to the same laws. The clockwork character of the Creation was not confined to celestial events but applied just as well to the phenomena on earth, notably, living organisms. The laws of mechanics, then, were valid also in the fields of medicine and biology. This belief received support from specific similarities between animal bodies and clocks and from spectacular achievements in the construction of automata.

The clock image called forth a number of related meanings. One sequence of associations, as just demonstrated, led from the clock's qualities of regularity, order, and harmony to the universe as a whole, to the Creation and the heavenly clockmaker, and to the mechanical character of all nature. Viewed from another angle, the clock image served to illustrate quite different notions.

Some viewers were not content to admire the clock's excellent qualities but were more interested in the machinery that had brought these qualities about. In comparing the clock with the planetary system, the living body, or the well-governed state, their focus was not on the order and harmony of these mechanisms' functioning but on the similarities in their structures. The purpose of such comparisons was to express approval and sometimes explicit delight over the thoughtfulness of the design, the perfect coordination of so many moving parts, and the astonishing efficiency of the total mechanism. Clockwork became a metaphor for the flawless working together of a complex combination of parts. It became one of the first concrete illustrations of a new abstraction that was in the process of formulation, namely, the concept *system*. Eventually the kinship was explicitly acknowledged. The article on system in Diderot's *Encyclopédie* adduced only one physical example: the mechanical clock.

The word *system* did not often appear explicitly in connection with clock metaphors. More commonly, when the meaning *system* was intended, that is, when the discourse was about a complex but well-integrated entity composed of many parts, the clock analogy was introduced directly. Systems illustrated by the image of the clock usually displayed certain common characteristics.

The actions of the various parts of a system always originated in the same way: they were initiated by a single central cause. The entire system was organized around a central authority that was directly linked to the multitude of organs or elements carrying out the system's various functions. The relationship between central authority and operating organ was conceived in principle as a cause-and-effect relationship, but it was usually described in terms of mechanical linkages. The custom of envisioning the system as ordered around a central authority led to certain rather obvious comparisons. God, the sun, the king, the brain (or the heart), and similar agents seen as central authorities were likened to the escapement, the program drum, the spring, or the weight of a clock. That the clock functioned by means of a central command structure was widely enough perceived, but there was no agreement as to which element in it was most analogous to the central authority. Was it its source of energy, its regulating element, or its memory device?

The relationship between the single central authority and the multitude of operating elements within the system had something hierarchical about it. Doubtless the system's organization was deeply shaped by the old and powerful intellectual tradition of arranging systems in hierarchies, many-layered and pyramid-shaped. But there were differences from the medieval concept of hierarchy: in the new clocklike systems there was much emphasis on the two ranks of the central authority and the operating elements; intermediate ranks between the highest and lowest, although not actually denied, received little attention. Emphasis was on the directness of the connection between these two ranks and on the indispensability for the whole of every single element of the system.

Indispensability of all parts did not imply equality in rank. Differences between the elements of the system were necessary because they were shaped and equipped differently according to their specialized tasks. Rigorous division of labor of the various parts gave the system efficiency. The specialties of the central authority were information, memory, judgment, and decision. The other elements of the system carried out their divers functions only as instructed by the central authority. No need was perceived, and no provision made, for return communication from the operatives to the center. No dialogue was possible between the center and the lower branches; the flow of communication was one-way—downward.

The similarity between the structure underlying this centralist and authoritarian conception of system and the structure of clocks and automata was acknowledged in numerous metaphors. This form of organization had its characteristic strengths and weaknesses and developed characteristic forms of behavior. A system that insisted on dealing with all problems centrally and that left its local operatives no discretion to act upon their own judgment was not well equipped to act spontaneously in the face of the unexpected. Its best hope for success was in the anticipation of all eventualities, in meeting them with plans and programs developed in advance, and in its dealings with the outside world, in always keeping the initiative. This was precisely how automata functioned; their mechanical programs—often interchangeable—enabled them to perform amazing feats, but they were incapable of making spontaneous responses to unforeseen challenges of even the simplest kind.

This preference for organizing complex systems around central authorities, for acting according to carefully laid plans, and for acting on one's own initiative was closely related to a deep-seated and widely shared belief in determinism. This was the belief that the world, once created by a maker of absolute and perfect wisdom, now proceeded along a course that was wholly predetermined by the choices made at the moment of creation. Such determinism was accurately illustrated in the way clocks functioned, be it in the automaton's preprogrammed feats or in the clock's basic function as timekeeper, where it duplicated the heavenly phenomena. Fittingly, the clock served as the conventional literary illustration of determinism.

These, then, were the principal implications of the clock metaphor. Recalling the original question, What were the metaphors actually saying? it is clear now that there is no single answer. Clock metaphors were saying several, albeit similar, things:

—They idealized the qualities of regularity, order, and harmony.

—They insisted on the clock as the prototype for the world, with regard to both its creation and its normal functioning.

—By pleading the mechanical character of the physical world, they sought to discredit magic; they sought to advance rationality both in the selection of evidence and in the analysis of causal connections.

—They promoted the mechanical clock as a physical illustration of the hitherto amorphous notion of *system,* that is, of an integrated assembly of numerous, dynamically interacting parts.

—They advertised the advantages of authoritarian, centralist command structures, be they in the body, in society, or in the universe.

—They illustrated and thus reinforced the general world view of determinism.

These statements summarize in condensed form but with all required fidelity to the original text the varying meanings of the vast majority of clock metaphors examined in the previous chapters. This is as far, however, as testimony of that time will take us. In any efforts to identify the single common denominator that unites all clock metaphors, the original sources will offer no further help. That question would not have been asked at the time. If all the clock metaphors had anything in common it would not have been anything explicit and distinct but something unexpressed and unacknowledged that existed only in the un- and subconscious minds of those who wrote clock metaphors and of those who applauded them.

Among the themes of the various groups of clock metaphors listed above, a certain basic affinity, some general connecting theme, is readily apparent. Together they outline a specific approach to the problems of establishing order among a mass of related, interacting parts and of organizing, maintaining, and controlling complex dynamic systems. The clock metaphor thus becomes an illustration of a general conception of order that is applicable to the most diverse areas of experience, from the living body to the state to the universe. The principal features of this authoritarian conception of order are its insistence upon control by one authority and a centralist command structure. The central authority communicates with the subordinate members of the system through rigid cause-and-effect relationships that are unidirectional and do not provide for or appreciate return signals ("back talk" in authoritarian usage, "feedback" in modern systems technology).

It is revealing to consider how the authoritarian conception of order envisions the maintenance of stability in its systems. The answer is contained in the image of the clock: an ideal clock is expected to run forever and function perfectly, entirely by virtue of its construction. A skillful designer knows the causes of all possible troubles; with such knowledge, he will fortify the mechanism against all dangers in advance and will program into it the appropriate responses to all foreseeable disturbances. In the face of unforeseen problems, however, the mechanism is helpless; the only salvation is with the intervention of its maker. The consequences of this approach to maintaining stability in the system are a pervasive dependence on central control and on decision making based on careful planning and forecasting and an abhorrence of on-the-spot decision making and of ad hoc problem solving, that is, of "muddling through."

What the clock metaphors meant but what they did not explicitly say can now be stated simply: they expressed, proudly and confidently, a peculiar conception of order that is perhaps best described as authoritarian and that was shared, if only unconsciously and in implicit form, by a significant part of the societies of early modern Europe.

Recognizing that the authoritarian conception of order was inseparably connected with the mechanical clock, one wonders how the emergence of the one was related to the invention and development of the other. This relationship, it seems, was one of intimate interaction and mutual reinforcement over three or four centuries.

When the mechanical clock was first invented, it was greeted—if writers like Dante and Suso are reliable witnesses—with almost religious veneration. Its immediate acceptance, rapid spreading, and quick technical maturing indicate how congenial and fascinating the clock seemed to the public. Demonstrating, in an impressively concrete manner, a particular kind of rationality and logic and a distinctive method of achieving desired results, it appealed to unexpressed desires and latent inclinations. For several centuries, the clock's most important function was perhaps to serve as an instrument of popular education and, indeed, indoctrination. To progress-minded Europeans of the Renaissance, the clock embodied the best things the future could bring: an end to magic and superstition, rationality in thought, and order in public life. No wonder that they pointed to this symbol of their aspirations at every opportunity.

While the authoritarian conception of order took shape in the minds of the literate upper classes, the clock also had its effect, partly in a nonverbal manner, upon the thinking and feeling of the unlettered rural majorities. They were not likely to get clocks into their hands, but they would see them in the village church, on the towers of the town, or at regional fairs. In both its roles, as a timekeeper and as a demonstration model of rational, purposeful action, the clock served as an important and purposely used instrument in preparing the masses for the ways of modern industrial society. This latter role of the clock—as an instrument for changing popular mentalities, attitudes, and behaviors—would warrant more detailed investigation.

The relationship between the technological advance of the mechanical clock and the formation and rise of the authoritarian conception of order was one of continuous interaction and mutual reinforcement. At first, as evidence of the public's general predisposition and sympathy for the new invention, the best minds and great material resources were committed to the development of the mechanism. In time the clock became more perfect and more widely accessible, and the more it affected public mentalities and attitudes, the more it generated support for itself. Thus a particular technology and a distinctive set of social ideals, values, and attitudes promoted each other in spiraling manner until the great climax in the seventeenth century with its extraordinary production of clocks and the conspicuous flourishing of the authoritarian conception of order.

We have perhaps finally arrived at an answer for our original question: Why the disregard for feedback before 1700? Was the public mind too fascinated with clockwork-like mechanisms, too full of the authoritarian conception of order, to be capable of any sympathy for another type of mechanism which, inconspicuous in appearance and subtle in functioning, represented a radically different and, indeed, rather anti-authoritarian conception of order?

6

Rejection of the Clock Metaphor in the Name of Liberty

Toward the end of the seventeenth century, when on the Continent the clock metaphor was at its height of popularity and influence, forces were gathering in England to oppose and defeat it. The immediate occasion arose from the design argument with its image of the Clockmaker God and the inherent dilemma of whether omniscience or omnipotence was to be viewed as God's overriding characteristic. The resulting controversy between intellectualists and voluntarists ended in Britain with the victory of the latter: if God's essential qualities were unlimited power and will, then the intellectualist doctrine of a predetermined creation was false, and the clock analogy for the world was no longer valid. In England, this judgment was widely accepted, notably by Newton and his circle. When Samuel Clarke, writing in 1704, rejected the clock metaphor in summary fashion, he doubtless represented the dominant view. And when he prevailed with this view in his widely publicized debate with Leibniz (his repeated claims that the clock image had nothing to do with either God or the Creation were left unanswered by Leibniz), it had the effect of a public certification of the clock metaphor's demise.[1]

The debate ended with Leibniz's death—Newton said, with evident gratification, that despair about losing the debate had killed him[2]—and the English side claimed victory. With regard to the whole of the debate which covered a complex of subjects, that claim is not undisputed, but with regard to the clock metaphor there could be no doubt. In England, the clock metaphor had lost all favor, and it was rarely used again except with negative connotations. The design argument itself, only in its voluntarist version, to be sure, survived longer although in a state of slow decline. It suffered particular damage when a philosopher of no lesser stature than David Hume constructed a painstaking and systematic refutation of it. In his *Dialogues concerning Natural Religion* (written in the 1750s, but published posthumously in 1779), he demonstrated that the world more resembled an organism than a human artifact, that the production of human artifacts and the creation of the world were things of different kinds, and that conclusions from human craft to the works of God were therefore not permissible.[3]

With its logical base thus compromised, the design argument nevertheless lived on. It achieved a last belated flowering in William Paley's *Natural Theology* (1802).[4] Paley was undaunted by Hume's theoretical refutation of the design argument; indeed he employed the clock metaphor more methodically than any previous English writer. His first two chapters were entirely devoted to the watch as an analogon of the Creation, and his analogy went into such detail that a set of illustrations for the *Natural Theology* drawn by James Paxton began with views of the parts of a watch.[5] Paley's book was not without influence. Both John Stuart Mill and Charles Darwin have testified to the formative impressions it made upon them in their youth. Darwin added that both the book as a whole and the design argument in particular had been strongly on his mind when he was working on the *Origins of Species,* which, of course, provided a definitive replacement for the notion that the world, like a complex clock, was created according to a comprehensive design.[6]

The clock metaphor's role in Paley's successful book, nevertheless, was an anachronism, an epilogue to a story that had ended almost a hundred years earlier. By the turn of the eighteenth century, the clock as a metaphor with positive connotations had been broadly and explicitly rejected in the English literature, and this rejection had an effect on the intellectual life of all of Europe. It is proper, therefore, to inquire into the causes of that rejection.

In the English attitudes to the mechanical clock, over the centuries, one can observe an intriguing ambivalence. The earliest mechanical clocks for which there are records were in England, and there is some probability that the clock was actually invented there. For the following three centuries, however, very little was heard of English clockmaking. Then, from the middle of the seventeenth to the end of the eighteenth centuries, Britain was Europe's undisputed leader in that technology. Even at the beginning, it seems, clocks in Britain were not received with unanimous delight. The denunciation of the new machine by the Welsh bard Dafydd ap Gwilym (ca. 1350), although singular, was extraordinarily passionate, and no English voices from the fourteenth and fifteenth centuries are known to have given such exuberant praise to the clock as did Dante, Suso, Froissart, Oresme, or Cusanus.[7] In the sixteenth century, English clock metaphors were ambivalent; this attitude persisted to the time of Shakespeare and longer. Characteristic qualities associated with the clock were quietness, truth, punctuality, and wholesome good health on the one hand, but coldness, gloom, and unpredictability on the other. Shakespeare's overriding emotion toward clocks was disgust ("clocks, the tongues of bawds"); his hostility was directed against their power to press the wide-ranging rhythms of our emotional lives into rigid, mechanical patterns. Shakespeare and other early seventeenth-century playwrights held up mechanical clocks as symbols of awkwardness, absurd complexity, and unreliability and, certain that those characteristics were foreign to the English character, identified such clocks as German.[8] A certain reserve toward things mechanical can even be detected in such contemporary natural philosophers as Bacon and Gilbert. The enthusiasm for clock imagery displayed later by mechanical philosophers like Robert Boyle and his disciples was of foreign inspiration: the mechanistic aspects of their philosophy were imported from Galileo, Mersenne, and Descartes. Boyle's infatuation with the mechanical clock seems to have been particularly linked with the great clock of Strasbourg. Even when the popularity of the mechanical philosophy was at its height, English enchantment with the clock image had definite limits; virtually no one accepted Descartes's animal automatism, and no one of consequence shared the Continental enthusiasm for rationalizing the state in the manner of clockwork. The most prominent English application of the clock metaphor was in the design argument, and even there it was comparatively short-lived.

A clue about the English ambivalence to clocks may come from observing a comparable ambivalence to things mechanical. The English words *engine* and *mechanical* (including their cognates) have traditionally had, apart from their familiar meaning, a negative connotation. *Engine* signified not only the kind of mechanical contrivance that it still does today but, consistent with its Latin ancestor *ingenium,* also meant *artifice, trickery, plot.* In the case of *machine* (originally *artful contrivance*), its ambiguity led to the formation of a new word: *machine* was retained for the neutral, purely technological sense in which it is understood today, while the negative connotations of trickery, intrigue, and deception became the burden of the word *machination.* This ambiguity of the English words *engine* and *machine,* which may be evidence of a deep-seated ancient distrust of technology, is also present, more or less, in their cognates in other European languages.

Peculiar to English, apparently, was the ambivalence of *mechanic* and *mechanical.* Besides their familiar value-free modern meaning, these words also had a pejorative sense. Samuel Johnson's *Dictionary* (1755) offered three definitions for "mechanick": "1. constructed by the laws of mechanicks; 2. skilled in mechanicks, bred to manual labor; 3. mean, servile, of mean occupation."[9] For the contemptuous use of *mechanical* in the sense of *vulgar, low, base,* the English literature of the sixteenth, seventeenth, and eighteenth centuries provides innumerable examples.[10] To the upper classes, *mechanical* signified the attributes, that is, the defects in courtesy, education, dress, and so forth, of the lowly who lived by physical labor. Accordingly, it served as an effective epithet for enemies of all kinds and was applied, for example, to partisans of the Copernican theory, members of the Royal Society, Oliver Cromwell, and Sir Isaac Newton.[11] *Mechanic* and *mechanical* were also pejorative in another, although related, sense. In this sense, *mechanical* described a characteristic somewhat indirectly linking machines and persons of low class, namely their lack of freedom. The *Oxford English Dictionary* lists examples for this connotation under the words "mechanic" (subheading A.5.b., "Involuntary, automatic"), "mechanical" (subheading A.4., "Of persons, their actions, etc.: Resembling [inanimate] machines or their operations; acting or performed without the exercise of thought or volition; lacking spontaneity or originality; machine-like; automatic"), and "mechanicism."[12]

There is even more direct evidence that *mechanical* was perceived as antithetical to *liberal*. A custom in the seventeenth and eighteenth centuries, dating from medieval scholasticism, contrasted the *liberal* arts with the *mechanical* arts. In the 1650s, for example, John Evelyn and others had been planning a *History of Arts Illiberal and Mechanical,* and Edward Gibbon, in his *Decline and Fall,* observed that in the household of a wealthy Roman senator there were represented "every profession, either liberal or mechanical."[13] This perceived contrast between the words *mechanical* and *liberal* perhaps offers a first small clue toward an understanding of the general British aversion to that archetypical mechanism, the clock.

This suggestion receives support from several other quarters. In the earlier English expressions of dislike of the clock, from Dafydd ap Gwilym to the seventeenth-century playwrights, the objections were focused on the clock's service as an instrument of regimentation and on its mindless, program-controlled, unfree manner of functioning. These objections emerge again in the debates about the design argument. The intellectualist version of the argument, which was strictly based on the notions of the Clockmaker God and the clockwork universe, was rejected for its determinism and its denial of free will even to God. Determinism, often illustrated with clock imagery, had not gone unopposed in Britain in previous centuries. Now, in the early eighteenth century, nothing would have been more fundamental to the prevailing world view than an uncompromising belief in the omnipotence and absolute free will of God. Samuel Clarke rejected "the notion of the world's being a great machine, going on without the interposition of God, as a clock continues to go without the assistance of a clockmaker" because it deprived God of his freedom to govern the world.[14] Similarly, George Berkeley suggested that "to suppose a clock is in respect of its artist what the world is in respect of its Creator" is to question God's omnipotence.[15] And Roger Cotes wrote in his preface to the second edition of Newton's *Principia,* "Without all doubt, this world, so diversified with that variety of forms and motives we find in it, could arise from nothing but the perfectly free will of God directing and presiding over all."[16] Those who believed in the free will of God also tended to claim free will for man. And whenever they did so, they were likely to express their rejection of the clock analogy. If man's mind functioned by the laws of mechanics, wrote Samuel Clarke in 1700, all thoughts would rigidly be determined by external causes, and "there could be no such thing in Us, as Liberty, or a Power of Self-determination. Now what Ends and Purposes of Religion, mere Clocks and Watches are capable of serving, needs no long and nice consideration."[17]

George Cheyne, who still employed the clock metaphor in the conventional, affirmative sense, came to the same conclusion, namely, that "that *Freedom* and *Liberty* of choosing or refusing which we find in ourselves, is altogether inconsistent with *Mechanism.*"[18] Berkeley (1732) reasoned similarly when he ridiculed the claims of some advocates of animal automatism,[19] and Colin Maclaurin (1748) finally pointed to free will as the very criterion distinguishing man from machine.

> The difference between a man and a machine does not consist only in sensation, and intelligence; but in his power of acting also. The balance for want of power cannot move at all, when the weights are equal: but a free agent, . . . when there appear two perfectly alike reasonable ways of acting, has still within itself a power of chusing.[20]

His archetypical mechanism was no longer the clock but, significantly, the balance.

In Britain, it seems then, the clock metaphor was rejected for precisely the same reason for which it was cherished on the Continent: as a symbol of authority, it was inevitably also a symbol of regimentation and oppression. The reason two closely related social orders took such contrasting attitudes toward the symbol of the clock was that they had embraced contrasting conceptions of order. The Continent, under the influence of its particular historical experience, had entrusted its salvation to the authoritarian conception of order described in the preceding chapter. The liberal alternative adopted in Britain was the consequence of the remarkable course of its recent national and social history.

Between 1588 and 1614, Great Britain, by its defeat of Spain, by the union of the English and the Scottish crowns, and by the subjugation of Ireland, had established itself as an integrated national state and as a leading European power. The drive to colonize North America and India was not only a symptom of an excess of expansionist energy but also a manifestation of the growth of certain characteristically British industries. Inseparably intertwined with this new emphasis on shipping, trading, and manufacturing was the rise, in size and status, of a commercial bourgeoisie. The emergence of this social class and the national concentration on commerce and industry together resulted in the establishment of a new economic system that proved remarkably vital and long-lived. Committed to the right of personal property and driven by the wish to increase such property, the participants in this system naturally cherished unrestrained economic competition and resented all outside interference with the freedom of the market. The unaccustomed anarchy of this newly emerging commercial society must have reminded Hobbes of man's "natural condition," which he described as a "war of all against all" in which "man is his own wolf," and in which "the life of man [is] solitary, poore, nasty, brutish, and short."[21]

Tensions between the traditional monarchy and the new economic and social forces led to a search for a more suitable political system. A variety of alternatives, from the intransigent absolutism of James I and Charles I to Cromwell's Commonwealth to the more moderate absolutism of the Restoration, were tested and rejected in quick succession. The system that finally won out as most suited to the new requirements was the constitutional or limited monarchy installed after the bloodless Glorious Revolution of 1688.

The century-long political struggle in Britain for a congenial political system was accompanied by an enormous output of political literature of every kind—from self-printed pamphlets to formal philosophical treatises. The struggle itself was perceived as a struggle for liberty; liberty was the dominant and recurring theme in this literature, the battle cry in attacking the old monarchical order and its recent absolutist reincarnation. "Experience . . . tells," wrote Thomas Sprat, historian of the Royal Society, "that greater things are produc'd by the *free* way, than the formal."[22] For a state that depended on its rapidly growing commerce, liberty was considered a prerequisite of survival. In the words of Lord Halifax, one of the architects of the Glorious Revolution, "We cannot subsist under a *Despotick* Power, our very Being would be destroyed by it; for . . . we are a little Spot in the Map of the World, and make a great Figure only by *Trade* which is the Creature of Liberty; one destroyed, the other falleth to the Ground by a natural Consequence."[23]

The wide range of benefits that liberty promised to bestow upon the state were advertised by chapter headings of Algernon Sidney's *Discourses on Government* (ca. 1680): "Liberty produceth virtue, order, and stability; slavery is accompanied with vice, weekness and misery." . . . "Mixed and popular governments preserve peace, and manage wars better than absolute monarchies." . . . "Men living under popular or mixed governments are more careful of the public good, than in absolute monarchies." But liberty was not only a source of benefits, it was also a human right: "The liberties of a nation are from God and nature, not from kings."[24] To John Milton, the poet of the Cromwellian revolution, the noblest result of a liberal government was intellectual freedom.[25]

Such expressions of commitment to freedom by leading Englishmen of the seventeenth century may serve here as a final commentary on the rejection of the clock metaphor. Liberty was the most frequently voiced slogan in the diverse revolutionary struggles of seventeenth-century England and also the unifying principle in the various changes that resulted. The English rejected the clock metaphor because they had clearly understood it to represent a world view, a value system, and a conception of order which were in direct opposition to their own.

The decline of the clock metaphor in England should be seen in its proper proportions. Rejecting the metaphor as a symbol for one's world view is not necessarily the same as resenting clocks as tools for practical tasks. In the middle of the seventeenth century, just when Robert Boyle began the massive use of clock imagery in his writings, Britain had assumed the leadership in European clockmaking. The so-called horological revolution that followed was primarily a British phenomenon. In mid-eighteenth century, when John Harrison provided the crowning achievement of the epoch, the ocean-going chronometer, the clock metaphor had long since fallen into disfavor. Metaphor and mechanism, it seems, existed on different planes, and their fortunes were no longer linked. A look at the new English clocks may make this observation understandable. Unlike the old German clocks, with their monumental architecture and bafflingly complex and diverse indications, the new English clocks, usually with a single dial in a wooden case, were startling in their austere simplicity. They were timekeepers, excellent ones in fact, and nothing else (see figure 1–8).

In Britain the clock metaphor had fallen victim to a new conception of order that could perhaps be described as liberal. Before turning to the substance of that new conception, it may be useful to survey briefly the clock metaphor's subsequent fate on the Continent.

Early in the eighteenth century, Continental Europe was learning to see Britain with new eyes. Through the centuries, it had regarded the island as a small and somewhat wild country at the fringes of civilization which neither attracted nor deserved much interest. Now the Continent discovered, to its astonishment, that Britain had grown into a world power: by trade and industry it had become the world's richest country, it had emerged from its recent political upheavals with an intriguing new political system, and, on top of all that, it was producing in such figures as Newton and Locke the leading thinkers of the age.

Visitors to the island who tried to make sense of so much good fortune were inclined to account for it in terms of one feature of English life that had impressed them the most, its liberty. After his visit to England in 1733, Voltaire wrote:

> The English are the only people on earth who have managed to prescribe limits to the power of kings by resisting them, and who by long endeavor have at last established that wise form of government in which the prince, all-powerful to do good, is restrained from doing evil; in which the nobles are great without insolence or feudal power, and the people take part in the government without disorder.

To explain the unusual mechanism by which the new system of government worked, he added, "The House of Lords and that of the Commons are the arbiters of the nation, and the king is umpire. This balance the Romans lacked."[26] He saw the foundation of the English political system in a characteristic interdependence of political liberty and a flourishing economy: "Commerce, which has brought wealth to the citizenry of England, has helped to make them free, and freedom has developed commerce in turn."[27] For Montesquieu, a visit to England from 1729 to 1731 proved formative. Convinced that the government of England was the only one on earth that regarded the maintenance of political liberty as the direct purpose of its constitution, he made the analysis of the English constitution with its characteristic separation of powers the centerpiece of his *Esprit des Lois* (1748).[28] Similar testimony from various European countries can be added.[29] And it can be shown that in the Continental popular mind the concept of liberty, in its many shadings, became deeply associated with the British nation in the course of the eighteenth century.

We may get an adequate indication of the impact of the liberal conception of order upon its Continental, authoritarian counterpart by sampling the subsequent fates of the clock metaphor in French and German writings.[30]

In France, the home of Voltaire and Montesquieu, the first Continental heralds of British liberty, the outcome of the encounter seems to have been inconclusive. The clock metaphor itself, as shown earlier, remained in use in the traditional manner without serious impairment. Liberty was cherished in the abstract, but the question of how to realize it in practical life proved troublesome. On the issue of political liberty, even Diderot and Rousseau, generally considered libertarians, took complicated, ambivalent positions. The ambiguity of the French enlightenment thinkers with regard to the ideal of freedom becomes more understandable when we examine their stands in the debate about determinism versus free will: virtually all of them sided with determinism.

Over this question even Voltaire had parted company with the idolized Isaac Newton: "Free-will is a word absolutely devoid of sense." He preferred the orderliness and predictability of mechanical as well as theological determinism:

> All things are machines merely; everything in the universe is subjected to eternal laws. Well, would you have everything rendered subject to a million caprices? Either all is the consequence of the nature of things, or, all is the effect of the eternal order of an absolute master; in both cases we are only wheels to the machine of the world.[31]

Representative figures like Diderot, La Mettrie, Holbach, and Rousseau similarly upheld determinism and rejected free will.[32] Diderot's position was somewhat complex: he understood that it was inconsistent to teach morals and to punish sin if one denied free will, and he opposed the most extreme form of determinism, fatalism, but when forced to take a stand, he sided with determinism against free will. "Study it closely and you will see that the word liberty is meaningless; there is none and there cannot be a free agent."[33]

That the materialists had no room for free will is not surprising. For La Mettrie, mental life was reducible to bodily processes: "Everything depends on the way our machine is running. One is sometimes inclined to say that the soul is situated in the stomach."[34]

The same beliefs, although less playfully stated, are laid down in materialism's grand manifesto, Holbach's *Système de la nature* (1770). With regard to the life of man, any freedom of choice is firmly and repeatedly denied: "In no one moment of his existence is man a free agent. . . . In man, free agency is nothing more than necessity contained within himself. . . . His [man's] life itself is nothing more than a long series, a succession of necessary and connected motion, which operates perpetual and continual changes in his machine."[35] Thus, man is fully subject to the all-embracing determinism of the universe of which he is such a small element.

In Germany, the debate about determinism took a different direction. Christian Wolff, whose most productive period had been the first quarter of the eighteenth century, was a radical determinist and an unsurpassed devotee of the clock metaphor. One of his immediate disciples, J. C. Gottsched, emulated him in these attitudes. By mid century, however, there was significant and steadily growing dissent.

The question, "Is the world a machine, or can it even be compared with one?" was discussed at length by Christian August Crusius (1715–75) and given an answer that differed considerably from Wolff's. Crusius defined a machine as a mechanical contrivance that, when combined with a power source, would execute "prescribed movements" (determinierte Bewegungen). The world, however, consisted not only of material things but also of spirits. Its course was not a series of predetermined movements but the result of the free actions of spirits: "It thus becomes obvious that the world is no machine and must not be compared with one." Moreover, the machine analogy was not only incorrect but also inappropriate because the world was something incomparably better and nobler, namely, the empire of God: "To be compared with a machine is degrading for the world."[36]

The case against determinism was also argued by the great mathematician Leonhard Euler (1707–83) in several of his *Letters to a German Princess* (particularly the eighty-third to eighty-seventh, written in 1760). His reasoning resembled that of Crusius but was more extensive and detailed. Specifically, it included attacks on Leibniz's principle of preestablished harmony, on the design argument, and on Wolff's mechanical determinism, all of which, to Euler, were inseparable from clockwork analogies. He began with a mildly sarcastic account of Leibniz's own illustration of the preestablished harmony between body and soul, that is, the image of two perfectly built clocks running in eternal synchronism although totally independent. Such, Leibniz had claimed, was the relationship between body and soul which, although incapable of interacting with each other, worked together flawlessly because of the perfection with which God had synchronized their actions in advance.

Euler rejected Leibniz's premise that body and soul cannot interact as "utterly destructive of human liberty." For "if the bodies of men are machines, similar to a watch, all their actions are necessary consequences of their construction."[37] Consequently, men could not be held responsible for their actions, and it would be absurd to punish criminals. With amusement he noted the embarrassment of determinists ("the article of liberty is the stumbling block of philosophy") who, on one hand, hated to deny human liberty but who, on the other, claimed that the human will was predetermined by its motives, "just as the motion of a ball on the billiard table is determined by the stroke impressed upon it, and that the actions of men are no more free than the motion of the ball."[38]

To resolve this dilemma, Euler proposed, much as Crusius had done, a distinction between "the nature of spirit and that of body."[39] The world of body, that is, of things material, when taken by itself, was indeed strictly determined, just as Wolff had taught. This world knew neither freedom nor responsibility.[40] But the world of the spirit was totally different: "Liberty, entirely excluded from the nature of body, is the essential portion of spirit to such a degree that without liberty a spirit could not exist; and this it is which renders it responsible for its actions."[41] This characteristic liberty of the spirit explained the inevitability of sin on earth.

The important truth, Euler insisted, was that the real world was made up of both bodies and spirits. If it consisted only of bodies, "the world would undoubtedly be . . . a mere machine, similar to a watch, which, once wound up, afterward produces all the motions by which we measure time."[42] All events of its career would have been predetermined from the moment of creation, like the program of an automaton.[43] The real world, however, consisted not only of bodies that obeyed the laws of mechanics but also the spirits whose wills were free and who were capable of acting upon bodies. Therefore, the universe was more than a machine and was "infinitely more worthy of the Almighty Creator who formed it."[44]

Such was, briefly, Euler's argument against determinism. Noteworthy is his extensive use of watch and clock imagery, especially his accurate identification of a music clock's program drum as the seat of that machine's deterministic behavior.

Our third and last witness on the clock metaphor's role in the debate of freedom versus determinism is Immanuel Kant. Kant's views are of interest not only because of his singular stature in the history of philosophy but also because of his well-known preoccupation with the problem of freedom. "Freedom," according to a modern expert, "is the central problem of Kant's entire work."[45] Kant himself had observed that "the principle of freedom . . . is the keystone of the whole architecture of the system of pure reason" and that "freedom is certainly the *ratio essendi* of the moral law."[46] Thinking about freedom meant coming to terms with the problem of determinism; in discussing deterministic behavior, Kant drew upon clock imagery in the familiar manner.

In his *Critique of Practical Reason* (1788), for example, clock and automaton were involved repeatedly in discussions of the relationship between necessity and freedom in nature. In one instance, Kant criticized the habit of determinists of bringing the concept of freedom into discussions about purely predetermined phenomena. One cannot claim, he argued, that a given crime resulted with mechanical inevitability from a set of given causes and at the same time talk of the free will of the criminal.[47]

When events took place in temporal sequence according to the laws of nature, that had to be acknowledged, Kant agreed, as a form of determinism. He called it the *mechanism of nature* and distinguished between an "*automaton materiale* when the machinery is impelled by matter, or, with Leibniz, *automaton spirtuale* when it is impelled by ideas."[48] If our freedom of will were only that of an *automaton spirituale,* he declared, it would be no better than a turnspit that, once wound up, will go through its motions mechanically.

Kant's preferred metaphor for determinism was generally the automaton. In discussing a proposition with deterministic implications, Kant concluded that in it "freedom could not be saved. Man would be a marionette or an automaton like Vaucanson's, fabricated and wound up by the Supreme Artist; self-consciousness would indeed make him a thinking automaton, but the consciousness of his spontaneity, if this is held to be freedom, would be a mere illusion."[49]

There is no need here to follow the complex argument by which Kant eventually reconciled the causality of physical nature with the principle of human freedom. But it is noteworthy that he rejected the notion of animal automatism in general. In his *Kritik der Urteilskraft* (1790), he demonstrated that organisms were not only essentially different but also of a higher order than machines: organisms had the ability to procreate whereas machines had only "motive force." In explaining the difference, his model of the machine was again the clock.[50]

Without following the debate about free will and determinism in German philosophy further, one can recognize a pattern in the course of the debate from Wolff to Kant that is similar to the British precedent. Direct influence from England could doubtless be documented. Christian Wolff, the uncompromising determinist, had held up the clock as the model for the world with all the reverence the clock had exacted at the height of its popularity. To Euler and Kant, half a century later, the clock was a commonplace mechanism that illustrated a rejected philosophic position.

Clock metaphors served not only in abstract debates but also in the discussion of more practical questions like that about freedom versus authority in government. The staunch determinism of the French philosophers of the Enlightenment had been hard to reconcile with a liberal political philosophy. Diderot, Holbach, Helvetius, Condillac, and others had agreed, with regret, that governments actually were and necessarily had to be machines. All they could do was counsel that the machine of government be kept as simple as possible. German philosophers shared this preference, but they did not think that simplicity alone was the answer. Kant warned of the dangers of despotism:

> Concerning the maintainance of justice in the state, the simplest [form of government] is certainly the best but, with regard to justice itself, also the most dangerous for the people in terms of the despotism that it invites. In the

> machinery of unifying the people by coercive laws, simplification is a sound maxim if all in the nation are passive and obedient to one who is above them; but subjects don't make citizens.[51]

The machine state, for all its efficiency, would inevitably become tyrannical.

Fichte postulated a direct correlation between the increase of mechanical elements in government and the loss of freedom. If a state were truly free,

> one wheel after another in the machine of the constitution of such a state would stop and be removed. . . . [The machine] would become increasingly simpler. Could the ultimate goal [the freedom of all] be ever completely achieved, then there would no longer be a need of a constitution; the machine would stop.[52]

The revolutionary tone of these quotations was not unusual for German political thinkers around the turn of the nineteenth century. Only a few years earlier, with the Prussia of Frederick II at the height of its successes, German writers had urged the clockwork state in uncompromising terms. But now—after Prussia's humiliating defeat, the French Revolution, the advances of liberalism (especially through Adam Smith's widely read *Wealth of Nations*), the emergence of the philosophy of German idealism, and the spreading of the Romantic movement—it was possible to criticize openly the old and to search for new political ideas. The absolutist machine state of the *ancien régime* which had failed so conspicuously was vigorously rejected. The Prussian disaster was blamed mainly on Frederick's "idea that a state is a machine constructed by the highest force." Largely responsible were Frederick himself and his "decrees begetting dead machines, which prove only the shortsightedness of this great despot."[53]

The shortcomings of the authoritarian form of government were often described with understanding and eloquence. Hegel charged that

> it is a basic premise of the new, only partly realized theories that a state is a machine with one single spring which imparts motion to all the other endless wheelwork. From the state's central authorities originate all institutions that establish the character of a society, and by them they are regulated, commanded, supervised, led. . . . This mechanistic, most intelligent and nobly intentioned hierarchy has no confidence in its citizens in anything; hence it cannot expect any in return.[54]

With his focus on the arrogance of the central authority, the subordination of the other elements of the system, and the inadequacy of communication between them, Hegel correctly diagnosed crucial weaknesses of the authoritarian concept of order.

Subsequent political thinkers, clearly disciples of Adam Smith, noticed other, subtler defects:

> Once the state is organized most perfectly like a machine, then the thing with its manifold wheels is supposed to run by itself, requiring only decisions from above, as from the hand of the machinist winding the clock.
>
> The more the idea of the machine state is practically realized, the more its apparent perfection generates the greatest unmitigated evils: it stifles individual thought and initiative which no drill master can teach, which no state can well do without, the stifling of which will haunt any state mercilessly in times of trouble.[55]

The neglected qualities, private enterprise and individual initiative, are, of course, precisely the qualities that liberal political systems needed to cultivate.

Another like-minded author attacked the mechanistic theoreticians of statecraft:

> To them state-building is a craft like organ building or clock making. . . . Specifying a mechanism and determining the weight that should set the machine in motion; wheelworks of institutions and social cooperation; the incentives of basic needs, or the stomach, hung on like a weight; and adding intelligence to the whole like a pendulum or correctional instrument—that is what they call a state. . . . But yet they have overlooked and left out the most important part. Each conceivable element of the state, each law, institution, etc. . . . has its own individual, mysterious life and its characteristic dynamics. The most exhaustive knowledge of these elements while in a state of dead rest means nothing. The student of statecraft must first return to ordinary reality, to experience. He must watch the law, the institutions, for a while in free life and in free motion; he must develop a feeling for the value and significance, and for the true functioning of the law which will count for more than the most thorough clockmaker's understanding of the matter.[56]

Such criticism of the machine state was common enough in German literature at the end of the eighteenth century, and it became conspicuously frequent in the early nineteenth century.[57] At the same time, subtle changes occurred in the use of mechanical imagery. *Machine* was no longer synonymous with *clock*. In some of the passages just cited, the clock was still expressly mentioned or, through references to wheels, gears, weights, and springs, clearly enough implied. Others invoked machines in the abstract. Occasionally there were specific references to other machines such as looms, mills, or pumps. As hostility to the concept *machine* mounted, however, perceptions of its character changed. Whereas the clocklike machine had connoted a tightly organized, complex unit running according to a rigid if intelligent plan, the new machine had grown in size and power and seemed to be characterized by overwhelming power and a capability of mindless, arbitrary violence.[58] The clock, while not regaining its former popularity, now seemed comparatively harmless.[59]

The more the machine symbolized evil, the more virtue became identified with nature and life. Hume's lighthearted saying of many years earlier—"The world plainly resembles more an animal or vegetable than it does a watch or a knitting loom"—was now the substance of a widely shared world view. As applied to political philosophy, the notion had a venerable precedent in the concept of the body politic. Organic imagery for the state had never lost its appeal, even when the clock metaphor had been at its peak; the two had coexisted peacefully to the point of forming mixed metaphors. Toward the end of the eighteenth century, when the clock and machine metaphors fell into disfavor, living organisms and the processes of nature were held up as models and symbols of good.[60]

"A monarchical state," wrote Kant, "corresponds to a living body when ruled by inherent laws of the people, and to a mere machine (e.g., a handmill) when ruled by a single absolute will."[61] Fichte called the "artificial political machine" of absolutist Europe a "strange work of art which, by its very composition, sins against nature."[62] Schiller, ostensibly describing the decline of the early republics of ancient Greece, lamented that "instead of rising up to a higher organic life, they sank down to a vulgar and crude mechanism, . . . an artful clockwork . . . where by piecing together infinitely many but lifeless parts a mechanical life . . . is formed."[63] To Hegel, organism and organization meant virtually the same thing, and both were equally antithetical to a machine: "But that rationalist state is not an organization but a machine, the people are not the organic body of a shared and rich life but an atomistic, lifeless plurality."[64] And Friedrich von Hardenberg, a scientist and romantic poet better known as Novalis, demanded that the old state, "this machine[,] be converted into a living, autonomous being by educating the citizens to public-spiritedness."[65]

This selection of samples from the discussions of the clockwork or machine state, from late eighteenth- and early nineteenth-century German writings, could be expanded at will and should give a sufficient idea of the course of the debate. Curiously, the idea of liberty was seldom cited in efforts to refute the mechanistic theory of government. The chief argument against that theory, instead, was the concept of the state as an organism, a concept associated with a set of values not easy to define but among which freedom was certainly not prominent.

II Liberal Systems

7

Imagery of Balance and Equilibrium

The first part of this study was concerned with interactions between the technology of clocks and automata and an authoritarian conception of order prevailing in early modern European society. In seventeenth- and early eighteenth-century Britain, this conception of order was rejected in the name of liberty and replaced by a radically different concept that, for the purposes of the following, will be termed *liberal.* This new liberal conception of order is the subject of the second part of this study, with special attention focused on its dependence on the emerging concept of self-regulation and its influence upon the new technology of feedback control.

The essential task of the liberal concept of order was to reconcile the conflicting values of freedom and order. That challenge could conceivably have been met within the old authoritarian structures by negotiating compromises case by case and by trading specific concessions in liberty and order. In actuality, the problem was solved far more ingeniously: a new general scheme, in theory as well as in practice, for the structuring of dynamic systems was discovered. That discovery was the realization that dynamic systems, in certain conditions, are capable of regulating themselves and of maintaining themselves in equilibrium by their own resources without the need of outside help, that is, without the intervention of a higher authority. In the following chapters, this capability will be called *self-regulation* to avoid the anachronistic term *feedback* to which it is equivalent. The notion of the *self-regulating system,* applicable to the most diverse fields, splendidly matched the needs of the liberal concept of order and was well on its way to broad popular acceptance in Britain by the mid-eighteenth century.

The clock metaphor had provided the authoritarian concept of order with a proven analog, indeed, a veritable working model of the construction and control of complex dynamic systems. If the clock was rejected as a mechanical paradigm because of its authoritarian character, was there a more liberal substitute for it? None was available; liberal thinkers, it seems, were unable to point to another concrete mechanism equivalent to but distinct from the clock that would spell out the kind of self-regulating system their principles and ideals demanded. No such mechanism, at any rate, was mentioned in contemporary discussions.

In antiauthoritarian writings, which often were also antimechanistic, metaphors of nature and organic life were invoked to illustrate and advance arguments. Such imagery reflected surprisingly little interest in any technical aspects of the functioning of the system. On the whole, organic imagery tended to be vague, unspecific, and rather emotional. It had little affinity with the concept of liberty and did not contribute to the formation of a liberal conception of order.[1]

The first practical applications of self-regulation to social systems, as they emerged in the eighteenth century, became known under such labels as the "balance of power," "checks and balances," and so forth. Their principal characteristic was the ability, alleged or real, to maintain automatically their own equilibrium. In studying their evolution, one notices at once that the image of balance has played a conspicuous role in the historical as well as logical formation of self-regulating social systems. And in probing deeper, one finds that the term *balance* quite generally was remarkably popular in English writings of the seventeenth and eighteenth centuries, far more so than on the Continent at the same time. To be sure, the stature of the balance was in no way comparable to that of the clock; as a metaphor it was old and familiar, as a concept elementary, as a mechanism trivial. Moreover, although the word balance may refer to a certain clock part, when used metaphorically in seventeenth- and eighteenth-century discussions, balance never had anything to do with clocks. The clock part called a balance has no semblance to the weighing instrument normally meant by "balance." Nor did the balance metaphor shape the liberal conception of order to the same extent that the clock had shaped the authoritarian one. The notion of balance is, however, a first clue, a scent to set us upon a trail.

The metaphor of the balance is as old as literature itself. It can be found in the writing of Pharaonic Egypt (late third millennium B.C.), in Homer's *Iliad,* in the Old Testament, and in Herodotus.[2] Since a balance is a tool for finding out which of two weights is heavier, it served in early literature as the metaphor of discrimination, decision, and judgment. The *mene tekel* of Belshazzar's feast—"Thou are weighed in the balance, and art found wanting" (Daniel 5:27)—is a classical specimen of this use.

An alternative aspect of the balance, which presented it not as an instrument of decision but as a model of equilibrium, emerged in classical antiquity. This view of the balance repeatedly served as an illustration of the various forces contending within political systems. There is a hint of it in Aristotle's *Politics:* to settle the eternal conflicts between the poor and the rich, the creation of a strong middle class, capable of *counterbalancing* the party that happens to be stronger, is urged.[3] Polybius developed this notion further in his theory of the mixed constitution: in the inevitable opposition between the monarchical and democratic elements of a state, he saw the aristocracy as capable of maintaining equilibrium by always siding with the weaker faction.[4]

The shift in the use of the metaphor perhaps indicated a new preference in values. The balanced state of a pair of scales had previously been deplored as undesirable, symbolizing indecision and confusion. Now equilibrium was seen as a positive value, as an antithesis to conflict, violence, and upheaval. These two contrasting views of the balance, as a symbol of either decision or equilibrium, have survived to modern times. Their rivalry can be observed in several different literary contexts, always following the same pattern: in the beginning the balance is invoked to illustrate difference and decision, but in the end it is praised as a symbol of equilibrium and harmony.

With such precedent in classical antiquity and by no means independent of it, the image of the balance came again into frequent literary use in sixteenth-century Europe. In subsequent centuries it would be applied to a wide range of social and political phenomena, in such combinations as "balance of power," "balance of property," "balance of wealth," "balance of parties," "balance of trade," "balance of payments," "checks and balances," and so forth. Not all of those phrases were of lasting significance, but three—balance of power, checks and balances, and balance of trade—became famous as labels of influential political and economic doctrines.

The notion of the balance of power derives from the empirical observation that even the most powerful state can be defeated by an alliance of its weaker neighbors. From this insight follows the rule that international equilibrium, that is, peace is best preserved if a given state makes its alliances with others always in such a way as to counterbalance the state that is individually most threatening. This principle was practiced with particular care by leaders of various Italian states in the Renaissance and found early expression in writings by Machiavelli and Guicciardini. In the sixteenth century, the rule or at least the imagery was invoked in attempts to curb the power of the House of Hapsburg. By the seventeenth century, the balance of power had become universally accepted as a proven rule of statecraft, rejected only occasionally by states with imperialistic designs, like the France of Louis XIV, which found itself opposed by alliances of weaker neighbors formed explicitly under the banner of the balance of power.[5]

It was in England, however, where the principle fell on the most fertile soil (the first recorded use of the term in English was in a 1579 translation of Guicciardini) and where it was to flourish with the most important consequences.[6] How did English authors envision the actual functioning of the balance? Earlier observers tended to see the states balanced in a static equilibrium. Sir Thomas Overbury (1609), for example, saw the whole of Europe stabilized in a combination of separate but occasionally interacting equilibria.[7]

It was a distinct historical experience—the successful foreign policies of Henry VIII and Elizabeth I—that had converted the British to the principle of the balance of power. English writers of the seventeenth century acknowledged this repeatedly. William Camden elaborated on the skill with which the principle had been handled by Queen Elizabeth,[8] and Francis Bacon praised the effectiveness of the balance of power by Elizabeth's predecessors.[9]

Proven by such successes, the principle in time advanced from a rule of practical statecraft to a divinely assigned national mission. Waller's *Panegyric to my Lord Protector* (1655) proclaimed:

> Heaven (that has placed this island to give law,
> To balance Europe, and her states to awe)
> In this conjunction does on Britain smile:
> The greatest leader, and the greatest isle![10]

The Tudor experience made a lasting impression, and the lesson was often reiterated by later statesmen.[11] The first English king to refer to the balance of power in a speech was William III, who exhorted Parliament on the eve of the War of the Spanish Succession (1701): "I will only add this—if you do in good earnest desire to see England hold the balance of Europe, it will appear by your right improving the present opportunity."[12] It was a war that Britain fought most deliberately for the maintenance of the balance of power in Europe. At the successful (for Britain) conclusion of the war, the principle was expressly incorporated into the Treaty of Utrecht (1713), which was dedicated to "ordering and stabilizing the peace and tranquility of the Christian world by a just balance of power" (ad firmandam stabilendamque pacem ac tranquillitatem christiani orbis, justo potentiae equilibrio).[13]

After the war, finding themselves elevated to the status of one of the three first, if not *the* first, powers of Europe, the British were inclined to give the balance of power the credit for this desirable outcome. Of the vast outpouring of tracts and pamphlets devoted to the subject, the following is characteristic (1720): "There is not, I believe, any doctrine in the law of nations, of more certain truth, of greater and more general importance to the prosperity of civil society, or that mankind has learnt at a dearer rate, than this of the balance of power."[14]

Although all agreed (at any rate in Britain, but later in the eighteenth century some Continental writers expressed dissent) that the balance of power was a good thing, there was less agreement as to what it meant. Was it an empirical fact or a theoretical ideal? The majority of writers treated it as a desirable state that could be brought about only by the skilled and purposeful actions of governments and statesmen. But there was also a temptation to view the balance of power as a natural process that produced equilibrium automatically, independently of, and if necessary in spite of deliberate human action, a temptation, that is, to view the balance of power as an example of a self-regulating process. This suggestion will be explored later.

The image of the balance of power was applied not only to the external relationships among sovereign states but also to the interplay of forces within a single country. Such balances had been suggested by Aristotle and Polybius, and Contarini (1543) had seen such a balance in the constitution of the republic of Venice: "This only cittee retayneth a princely sovereigntie, a government of the nobilitie, and a popular authority, so that the formes of all seem to be equally balanced, as it were with a paire of weights."[15] He was expressing high praise, for the political system of Venice that he saw realized the ideal of mixed government advocated by Polybius.

City republics in medieval and Renaissance Europe had occasionally and temporarily approached this form of government, but the establishment of the first successful constitutional government over a large state in modern Europe was an English achievement. It was borne from the tensions after Queen Elizabeth's death between the Stuart monarchy and its aristocratic supporters, on the one hand, and the rising middle class, on the other. In time, social polarization turned into political conflict that, after continuing for much of the seventeenth century, ended with the Glorious Revolution of 1688.

It could perhaps be shown that the new constitutional monarchy then instituted in Britain had been consciously intended and deliberately designed by its architects as a self-regulating system. Such a demonstration would require a detailed review of seventeenth-century English constitutional history, which would clearly be impracticable here. Instead, we will only review or at least sample the mechanical imagery employed in the debates of the period in order to learn whether and to what extent the notion of a self-regulating system of government was spelled out in mechanical terms.

Those who had fought for and won this revolutionary form of government, that is, members of the new propertied middle class represented in Parliament by the House of Commons did not wish to be viewed as rebels. They preferred to present themselves as conservatives defending ancient inherited rights against the encroachments of an aggressive absolutist monarchy. They insisted that their innovative constitutional ideas were part of old and venerable traditions and pointed to Polybius as the author of the concept of mixed government, to a "Gothic balance" in the mythical Anglo-Saxon past as an ideal combination of monarchy and democracy, and to the Magna Charta as the beginning of a limited monarchy.[16] Belief in these traditions had even been shared to some extent by their monarchist opponents. In the course of the unsuccessful negotiations that preceded the civil war, King Charles I, in a document known as *The King's Answer to the Nineteen Propositions,* had publicly declared himself in favor of the principle of mixed government:

> There being three kinds of government among men, absolute monarchy, aristocracy, and democracy, and all these having their particular conveniences and inconveniences, the experience and wisdom of your ancestors hath so moulded this out of a mixture of these as to give to this kingdom (as far as human prudence can provide) the conveniences of all three, without the inconveniences of any one, as long as the balance hangs even between the three estates.[17]

This statement, much quoted during the subsequent disputes because it represented a widely acceptable intermediate position, contains three important elements: it pays proper tribute to tradition; it concedes three coequal branches of government (a notion that in time would evolve into the concept of the separation of powers); and it acknowledges the need for equilibrium among the three branches. The character of this equilibrium was not specified. Since it did not assert the opposite, that is, it did not suggest that equilibrium was to be maintained forcibly by some definite agent, one may interpret this as a hint of a self-balancing equilibrium.

In English political debates of the next half century, balance imagery, as applied to the branches of government or to diverse social groups, was used frequently; with regard to the process that was to establish equilibrium, none were too clear. Under Cromwell, the Parliament used language remarkably similar to the king's. Acting somewhat as its ideological spokesman, John Milton demanded (ca. 1650) that "the balance therefore must be exactly so set, as to preserve and keep up due authority on either side, as well as in the senate as in the people [i.e., the houses of Lords and Commons]."[18] On another occasion, while acknowledging its indebtedness to classical antiquity, he praised the internal balance as one of the chief distinctions of the English political order:

> There is no civil government that hath been known, no not the Spartan, not the Roman, though both for this respect so much praised by the wise Polybius, more divinely and harmoniously tuned, more equally balanced as it were by the hand and scale of justice, than is the Commonwealth of England; where, under a free and untutored monarch, the noblest, worthiest, and most prudent men, with full approbation and suffrage of the people, have in their power the supreme and final determination of highest affairs.[19]

Thus there was much agreement between the two warring sides on the advantages of mixed government, on the desirability of a balance among the social forces in the nation, and on the general value of equilibrium. Why did such agreement not lead to reconciliation? Among other reasons, one was perhaps that the mechanism for establishing equilibrium had not yet been worked out. It was not clear whether the equilibrium had to be established by the force of the victor (which is what both opposing parties tacitly assumed) or whether it would come about automatically. In the latter case, a more complex model than the simple balance would have been needed to illustrate the process. Lacking this, the balance continued to be popular as a rhetorical device without operational meaning. As such, it was invoked often.

Cromwell himself was fond of the image and repeatedly referred to the "true and equal balance" established by government or to the "need of a check or balancing power" between the two houses of Parliament.[20] An author who used the word balance with extravagant frequency was James Harrington, a staunch admirer of Cromwell who, in his *Oceana* (1656), tried to construct a theoretical foundation for the Commonwealth. Harrington advocated a republic where the land and hence the power (he dismissed the importance of other forms of wealth) would be held by armed freeholders: "But if the whole people be landlords, or hold the land so divided among them that no one man or number of men within the compass of the few or aristocracy overbalance them, it is a commonwealth."[21] This sentence offers a sample of Harrington's use of the word balance; alone, in combination (e.g., "balance of the king," "provincial balance,") or in the form "overbalance," the word appears in the *Oceana* hundreds of times. Used loosely and imprecisely, it virtually never signified equilibrium or equality but normally had the meaning of overbalance, that is, of preponderance or predominance. One will therefore look in vain in Harrington's writing for balance imagery illustrating equilibrium. Harrington's *Oceana* is best understood as an opposition document directed against the concept of mixed or limited government with its characteristic balance of three separate powers, an idea that, since the *King's Answer* of 1642, had become the centerpiece of a consensus that was to guide British political life for centuries.[22] It is curious but not unfitting that, in the thirty years following the appearance of *Oceana,* that is, during the Restoration, no notable discussions of the nature of equilibrium in government nor at least further samples of balance imagery could be found. There are none even in John Locke's *Two Treatises of Government* with its famous elaboration of the separation of powers. After 1688, however, the use of such imagery was resumed with unprecedented gusto.

The phrase, balance of trade, like the "balance of power" in its several meanings, seems to have its origins in sixteenth-century Italy, adapted probably from the terminology of bookkeeping and accounting, which had matured there in previous centuries. The expression first appeared in English in 1615 when a customs official called the comparison of the totals of merchandise imported to and exported from England in a year the balance of trade (the concept itself, under other names, had been familiar in the sixteenth century and was not unknown even in the fourteenth).[23]

In print, the term appeared first in 1623 in the title of a book by Edward Misselden, *The Circle of Commerce, or the Ballance of Trade,* which offered a detailed definition:

> For as a paire of Scales or Ballance, is an Invention to shew us the waight of things, whereby we may discerne the heavy from the light, and how one thing differeth from another in the Scale of Waight: So is also this *Ballance of Trade,* an excellent and politique Invention, to shew us the difference of waight in the *Commerce* of one Kingdome with another: that is, whether the Native Commodities exported, and all the forraine Commodities Imported, doe ballance or overballance one another in the *Scale of Commerce.*[24]

The image of the balance was seen here in its older aspect, as an illustration of comparison or difference rather than as one of equilibrium or equality. And the image led to the appropriate conclusions. In 1616 Francis Bacon included in a letter to the Duke of Buckingham, a favorite of King James I, this admonition:

> The kingdom is much enriched of late years by the trade of merchandise which the English drive [*sic*] in foreign parts; and, if it is wisely managed, it must of necessity much increase the wealth thereof: care being taken, that the exportation exceed in value the importation: for then the balance of trade must of necessity be in coin or bullion.[25]

The conviction expressed here—that a positive or favorable balance of trade was a good thing for a nation's economy—came to be widely shared for many years. It became the centerpiece of the doctrine of mercantilism, which controlled the economic policies of most European states in the seventeenth and eighteenth centuries. Commanded by this doctrine, governments did their best to increase exports, decrease imports, and achieve a net inflow of cash. They sponsored labor-intensive industries, favored the import of raw materials and the export of finished industrial products, and discouraged the opposite. "It is therefore a general maxim to discourage the importation of work, and to encourage the exportation of it" wrote Sir James Steuart in 1767.[26] In the quest for the positive trade balance, governments imposed trade restrictions and tariffs; they tried to subordinate whole national economies to central planning and operated certain industries themselves. The mercantilist doctrine, in short, relied on authoritarian methods. It was congenial to absolutist governments and helped to confirm them in their authoritarian centralism.

With connotations such as these, the term *balance of trade* was widely used in the seventeenth century and has remained popular ever since. But what did it have to do with the rise of the habit of thinking in terms of equilibrium and of the notion of the self-regulating system? The answer is that as, in the late seventeenth and early eighteenth centuries, the simple and straightforward doctrine of mercantilism was subjected to subtler analysis, the balance of trade began to be seen in a different aspect. Earlier the balance had been judged to be in its proper state when one scale—the one that measured the flow of payments into the country—was unambiguously low. Now the preferred state of the balance was equilibrium, preferred not for its inherent advantages but for its recognized inevitability.

8

Attraction and Repulsion

A self-regulating system may be pictured as a balance connected with some mechanism that, as soon as one scale drops, begins to remove weight from that scale and add it to the other until equilibrium is restored. It is likely that the fascination with self-regulating systems which took hold in Britain around the turn of the eighteenth century had its origins in the widespread popularity of the balance image. Another explanation for this fascination has also been offered, however, by former U.S. president Woodrow Wilson: "The government of the United States was constructed upon the Whig theory of political dynamics, which was a sort of unconscious copy of the Newtonian theory of the universe."[1] This suggestion is not unreasonable because the Newtonian system of orbital mechanics, which accounted for the characteristic motions of planets, satellites, and comets through a dynamic equilibrium between gravitational and inertial forces, might perhaps be viewed as self-regulating. The question is whether what is compatible with reason did indeed take place in history. We must investigate how the idea was formulated by Newton himself; how it, in its original astronomical context, was received by the public; and whether and how it was transferred as metaphor to other subject matters.

The early origins of the idea that planetary motion is an equilibrium of two opposing forces, which included contributions from Jeremiah Horrocks, Giovanni Borelli, Christian Huygens, and Robert Hooke, need not be considered here because they were contained in technical discussions of specialists which did not reach public attention.[2] Newton's earlier formulations of the idea were expressed only in private correspondence. He wrote, for example, to Robert Hooke on 13 December 1679: "I agree with you . . . that if [an orbiting body's] gravity be supposed uniform it will not descend in a spiral to the very center but circulate with an alternate ascent & descent made by it's *vis centrifuga* & gravity alternately overballancing one another."[3] A little later (April

1681?), he referred to the same dynamic scheme in the somewhat different physical context of cometary travel. He urged the view that

> ye comet to be directed by ye Sun's magnetism as well as attracted, & consequently to have been attracted all the time of its motion, as well in its recess from ye Sun as in it's access towards him, & thereby to have been as much retarded in his recess as accelerated in his access. & by this continuall attraction to have been made to fetch a compass about the Sun . . . , the *vis centifuga* at [perihelion] overpow'ring the attraction & forcing the Comet there not withstanding the attraction, to begin to recede from ye Sun.[4]

These brief statements suggest some form of dynamic equilibrium: as the planet, traveling on its elliptic path, gravitates toward the sun, it gains speed; this speed, in turn, allows it to overcome the gravitational pull and move away from the sun. In so doing it is decelerated until it reverses its course again toward the sun to repeat the cycle. The statements might have been vivid enough to kindle in the contemporary reader an interest in dynamic equilibrium as a more general phenomenon. They were made in private letters, however, and had only a small audience.

Publicly, in his *Philosophiae naturalis principia mathematica* (1687), Newton handled the subject quite differently. In accord with the book's title, he described the dynamic behavior of celestial bodies mathematically, not physically. The minimum of physical description offered at the beginning seems almost studiedly undramatic. The notion of centrifugal force was given up, and all that was left of the struggle of opposing forces was expressed in the definition of one of the book's basic concepts, *centripetal force:*

> Of this sort is gravity, . . . magnetism, . . . and that force, whatever it is, by which the planets are continually drawn aside from the rectilinear motions, which otherwise they would pursue, and made to revolve in curvilinear orbits. [Orbiting bodies] all endeavor to recede from the centres of their orbits; and were it not for the opposition of a contrary force which restrains them to, and detains them in their orbits, which I therefore call centripetal, would fly off in right lines, with an uniform motion.[5]

And he went on to describe the various trajectories of bodies in gravitational space by introducing the thought experiment of a horizontally aimed cannon on a mountain top whose projectile, depending on the strength of the charge, would either fall down in a parabolic path, circle the earth in circular or elliptical orbits, or escape altogether from the earth's gravity.

In principle, Newton's method consisted of counting the centripetal forces against the inertial forces, as defined by his first law of motion. He applied this procedure purely mathematically to a great variety of configurations of celestial bodies, achieving splendid agreement with empirical observation and thus proving the correctness of his theory of the universe. The discussion was confined to planetary motion at specific points along the orbit. The complexities of the concept of dynamic equilibrium were not considered; presumably Newton was only too well aware of the mathematical difficulties involved and of the inadequacy of the mathematical tools available to him. Moreover, in view of the perturbating forces of other planets, he had doubts about the long-term stability of systems of orbiting planets and believed in the need for occasional divine interventions to restore equilibrium.[6]

Newton's *Principia,* for all its fame, was too difficult a book to find many readers. A large literature sprang up in the following decades to make the Newtonian philosophy accessible to a larger public. How were the phenomena of orbital dynamics with the attendant problems of dynamic equilibrium handled in that literature? Judging by a sampling of some two dozen representative works, the Newtonian popularizers, when it came to the dynamics of orbiting bodies, followed closely the original's bland style and its caution in the use of imagery. Formulations that drew attention to the concept of equilibrium or that hinted at problems of instability were rare. The following samplings are representative of the stronger formulations offered in this literature.

John Harris, in his *Lexicon Technicum* (1708), dealt with the matter by citing an account of the motions of planets from Whiston's *New Theory of the Earth* (1696): "From the uniform Projectile Motion of Bodies in Straight Lines, and the universal Power of Attraction or Gravitation, the Curvilinear Motion of all the Heavenly Bodies do's arise." And the word *equilibrium* was actually used:

> When the Projectile Motion of the *Planets* is in its Direction perpendicular to a Line from the Sun, and in its Degree of Velocity, so nicely adapted and contemper'd to the Quantity of the Sun's Attraction there, that neither can overcome the other, (the Force of Gravitation towards the Sun, and the Celerity of the Planets proper Motions, being perfectly *in aequilibrio*) the Orbits of such Revolving Planets will be Compleat Circles.[7]

Somewhat more forceful language is found in William Derham's *Astro-Theology* (1715) under the heading, "Of the Power and Usefulness of Gravity to retain the Planets within their Orbits." A description of how the interplay between gravity or "attractive force" and centrifugal or "projectile imprest" force kept the planets on their paths around the sun concluded with this sentence: "We have another exquisite Nicety in the works of the Creation, that justly deserves the greatest admiration and praise; that among so many immense moving Masses, there is not one without this exquisite Aequilibration."[8]

Another witness, J. T. Desaguliers, in his *Physico-Mechanical Lectures* (1717), applied the concept of centrifugal force to the phenomenon of orbital travel, illustrating it with Newton's picture of the cannon on the mountain top:

> If a Cannon-Ball was shot Horizontally (or upon the Level) from a Mountain, with a Force of Powder equal to the Earth's attraction, it wou'd constantly go round, and never come to the Ground; like a Planet. The *Projectile Motion,* or Centrifugal Force arising from it, wou'd keep it from falling to the Earth, and the Attraction of the Earth, wou'd keep it from flying out of its Orbit, and quitting the Earth by moving off in the Tangent.[9]

In his account of the planetary orbits, Desaguliers described a balance of opposing forces in a similar manner.[10]

These statements deal with their subject matter clearly enough, but they do not promote dynamic equilibrium or self-regulation as general concepts of importance in their own right. Someone whose fascination with such concepts was already well established would doubtless have been pleased at the demonstration and confirmation provided by Newton's system. But it is difficult to believe that discussions of the kind quoted (and I doubt that more compelling formulations are to be found in the post-Newtonian literature) would have kindled a fascination for these concepts in readers with unprepared minds.

The Newtonian philosophy or what was popularly taken for it may, however, have reinforced equilibrium thinking in another way. Popularizers, as they attempted to make Newton's system comprehensible to larger segments of the public in special treatments prepared for "gentlemen," ladies, adolescents, and so forth, tried to reduce it to a few fundamental principles that were easy to grasp.[11] One of the notions employed for this purpose, and perhaps the most popular one, was *attraction.* Attraction comprised such phenomena as gravitation, magnetism, static electricity, and chemical bonding. "Keill," said one popularizer of another, "*makes use of three Principles,* viz. 1. Empty Space. 2. The infinite Divisibility of Quantity. 3. The Attraction of Matter. *And affirms, that all* Physics *depends thereon.*"[12] For himself the same author announced:

> It seems not consistent with the Regard a Philosopher should have to the Uniformity of Nature, every where observable, to call in a new Principle at every knotty Point. Those which I make use of are,
>
> First, *Attraction of Gravitation.* That is, a Disposition in Bodies to move towards each other, even when at great Distances asunder.
>
> Secondly, *Attraction of Cohesion.* That is, a like Disposition in Bodies to move towards each other, but distinct from the former, in as much as it is observed to take Place only when the Bodies are very near together.

> Thirdly, *Repulsion,* or a Disposition in Bodies, whereby in some Cases they endeavour to avoid, or fly from each other.[13]

What was only suggested here was given radical expression in a book by Gowin Knight, a respected Fellow of the Royal Society and first librarian of the British Museum; it had the title, *An Attempt to Demonstrate, that all the Phaenomena in Nature May be explained by Two Simple Active Principles, Attraction and Repulsion* (1748).[14] Even Newton himself had hinted that he saw in the contrasting qualities of attraction and repulsion a general principle that might contribute to the finding of a unified theory for both micro- and macroscopic phenomena.[15] In that sense "attraction and repulsion" had enormous appeal and became one of the favorite clichés of Newtonianism. Scientists tried to use this pair of opposites to explain a wide range of physical phenomena, and writers of all kinds wove them into their poetry and prose. These two words clearly met the popular taste.

Whoever spoke of attraction and repulsion was implying that reality lay somewhere between the two and that their mutual opposition tended to produce equilibrium. Attraction and repulsion then simply became new names for the two scales of the old balance. And in some ways the new imagery was not an improvement. Whereas the old picture of the balance was hard and concrete and serviceable in the building of sound models and analogies, the concepts of attraction and repulsion were abstract in the first place, metaphors themselves when applied to the behavior of celestial bodies, and were incapable of being translated into logical, mathematical, or mechanical procedures when applied to anything else. In consequence, it is hard, despite the frequency of loose references to attraction and repulsion, to locate clear statements that relate the two to the notion of equilibrium.

Newtonian imagery, both in the narrow sense of planetary dynamics and the broad sense of attraction and repulsion, was occasionally invoked in discussions of social phenomena. Specimens of the first kind are not impressive. When Walter Moyle (ca. 1699) said that "each branch [of government] was a check upon the other, so that not one of them could exceed its just bounds, but was kept within the sphere in which it was circumscribed by the original frame,"[16] it is uncertain whether he had the Newtonian system in mind. A passage from Bolingbroke (1730s) is only a little more specific:

> A king of Britain is . . . what kings should always be, a member, but the supreme member . . . of a political body: . . . he can move no longer in another orbit from his people, and, like some superior planet, attract, repel, influence, and direct their motions by his own. He and they are parts of the same system, intimately joined and cooperating together, acting and acted

> upon, limiting and limited, controlling and controlled by one another; and when he ceases to stand in this relation to them, he ceases to stand in any.[17]

Adam Smith finally described the natural price of a commodity as "the central price, to which the prices of all commodities are continually gravitating."[18]

Quotations involving gravity in the looser sense are only marginally more rewarding. The most extravagant comes from one of Newton's disciples, J. T. Desaguliers who, for the coronation of King George II in 1728, presented an allegorical poem entitled *The Newtonian System of the World: The Best Model of Government*. Hyperbolically, as the occasion demanded, Desaguliers subordinated both the British kingdom and the physical world under the principle of gravitation:

> The limited Monarchy, whereby our Liberties, Rights, and Privileges are so well secured to us, as to make us happier than all the Nations round about us, seems to be a lively Image of our System; and the Happiness that we enjoy under His present MAJESTY'S Government, makes us sensible, that ATTRACTION is now as universal in the Political, as in the Philosophical World.[19]

Giving due credit to Newton for this achievement:

> By *Newton's* help, 'tis evidently seen
> Attraction governs all the World's Machine,

the poem ends with the lines:

> Attraction now in all the Realm is seen
> To bless the Reign of George and Caroline.[20]

Since it was written by a physicist, the piece cannot be taken as representative of contemporary political thought, nor did it have or deserve wide circulation. Nevertheless, such sentiments were shared by others. Samuel Bowden, in a eulogistic poem on the death of Newton, asserted that Newton "saw perpetual *Gravity* obtain, And o'er the System hold coercive Reign," and compared the relationship between the two forces ("projectile" and "central") that control the movement of a celestial body to the "Balance of Empire."[21]

The thought was again expressed in an anonymous pamphlet of 1758: "What gravity or attraction, we are told, is to the system of the universe, that the ballance of power is to Europe."[22] Later in the century when he was discussing the internal mechanism of a state, Edmund Burke referred to "that opposition of interests, . . . that action and counter action, which, in the natural and in the political world, from the reciprocal struggles of discordant powers, draws out the harmony of the universe."[23] Statements of this kind can also be found in the works of some French authors of the latter eighteenth century such as Montesquieu, Holbach, and Helvetius.[24]

There can be no doubt that equilibrium thinking in general, and the concept of dynamic equilibrium in particular, played certain roles in the Newtonian philosophy. These concepts were employed in the scientific literature with adequate clarity but without didactic emphasis. To some extent the imagery of orbiting planets and of attraction and repulsion was applied to social systems, but such applications were few in number and weak in emphasis. It seems unlikely that the remarkable popularity of concepts and systems involving equilibrium in early eighteenth-century Britain was solely a consequence of Newtonian ideas. It is more likely that the occurrence of such notions as balance and equilibrium among the Newtonian ideas was a reflection of the growing popularity of equilibrium thinking at the time.

9

Self-balancing Political Systems

The image of the balance in equilibrium symbolized a condition that was held to be highly desirable. The question now was only, how was this condition to be established and maintained as widely as possible in practical life? The initial response had been simple and direct: equilibrium had to be achieved by purposeful, judicious action. If the weight in one scale was heavier than that in the other, then some responsible person had to make the appropriate adjustments. Establishing and maintaining the balances of power, government, or trade was the work of rulers and statesmen; the equilibrium of the cosmos, in the view of Sir Isaac Newton, depended on active interventions of God.

From around the turn of the eighteenth century on, a new answer to the question about the maintenance of equilibrium was heard more and more often. It was suggested that, in certain circumstances, equilibrium was capable of maintaining itself automatically, without external help. Thus, dynamic systems could be constructed in such a way as to be self-balancing or self-regulating. Outside interference in such systems was unnecessary and undesirable; equilibrium could not, and should not, be imposed by external agents. To some extent, perhaps, this new concept of equilibrium had been suggested by practical experiences. It was extremely congenial to a liberal mentality; a system that could balance and regulate itself in full autonomy and independence, without help from a higher authority, would be the foundation of a liberal form of order.

The concept of a self-regulating system was a subtle and abstract notion, a totally artificial mental construct. It took several decades of discussion to produce clear descriptions and a critical understanding. Concurrently, in an inarticulated, unconscious manner, the concept gradually entered popular usage. In time it was applied, casually or seriously, to the most varied physical and social phenomena. The following will survey such efforts in three areas already introduced, namely, those associated with the terms *balance of power, balanced government,* and *balance of trade.* The outcomes were uneven: in the first it was open failure, in the second a success that was conditional and not undisputed, and in the last a sweeping triumph.

For the balance of power, the story is quickly told. By the beginning of the eighteenth century, this balance was generally accepted as a sound principle of foreign policy; Britain felt a special affinity with it, for Britain's newly acquired power over the affairs of Continental Europe was directly attributable to this principle. Disagreement continued only over the question of how exactly the balance of power was maintained. It was clear now that earlier interpretations like those of Sir Thomas Overbury, envisioning a static balance that, once established, would last forever, were inadequate.[1]

Subtler observers, especially if they had an open mind for the lessons of practical politics, knew that in international affairs the distribution of power was in constant flux and could not be balanced once and for all. Balancing the power of sovereign nations, one anonymous critic in 1712 observed, was like equalizing the wealth of individuals:

> And thus it is among Nations, it is impossible to bring them to an equal Ballance of Power or Riches. And if it were done, if all the Nations of the Earth were reduced to an equal Ballance even of a Grain Weight, then a Grain on any Side would cast the Ballance. And this Ten Thousand Accidents every Day would produce, a prosperous Voyage on one Side, and Unfortunate on another; a Wiser or a Weaker Administration in one Government than another, would turn the Ballance vastly, So that We must Ballance the Wisdom, the Industry, and the Courage of Men, as well as their Honesty and Conscience; and likewise secure Providence not to favour one more than another, if we would fix the Peace of the World upon this Project of Ballancing.[2]

The author sounded pessimistic because he did not know how to reconcile his static conception of the balance with his awareness of the constantly changing forces in actual politics.

Lord Bolingbroke joined in this diagnosis, but for him, as a seasoned politician, the therapy could not be in doubt; governments must constantly watch the balance and be ready to take corrective action when required:

> The scales of the balance of power will never be exactly poised, nor is the precise point of equality either discernable or necessary to be discerned. It is sufficient in this, as in other human affairs, that the deviation be not too great. Some there will always be. A constant attention to these deviations is therefore necessary.[3]

Bolingbroke expressed a majority view; the balance of power was nothing more complicated than a general guideline of political action, and it was not automatic.

If a shrewd political mind like Bolingbroke was unable to view the balance of power as self-balancing, it was not for lack of imagination. He had presumably considered the possibility and, in the light of practical experience, rejected it. A few voices that were more naïve suggested the idea, only to be universally ignored. For example, the idea of a self-regulating balance of power is implied, although none too clearly, by the anonymous English pamphleteer of 1758: "What gravity or attraction, we are told, is to the system of the universe, that the ballance of power is to Europe: a thing we cannot just point out to ocular inspection, and see or handle; but which is as real in its existence, and as sensible in its effects, as the weight is in scales."[4] The author accounted for the characteristics of self-regulation not through the design of the system but through some immanent, ineffable quality comparable to gravity. Newton would probably have preferred to be left out of that discussion.

A strong, unequivocal claim that the balance of power was self-balancing came from a non-Englishman, Jean Jacques Rousseau in 1761:

> But whether we pay attention thereto or not, this equipoise subsists without the interposition of a second person, and wants no foreign effort for its preservation; and when an alteration is made on one side, it immediately re-establishes itself on the other: so that if those princes, who are accused of aspiring to universal monarchy, have really aspired thereto, they have manifested more ambition than genius, for in what manner can they have imagined to themselves that project, without perceiving the ridicule thereof in the first moment? . . . In fine, not one amongst them being able to possess resources which the others cannot acquire, resistance at length becomes equal to the effort, and time soon re-establishes the rapid accidents of fortune, if not for each particular prince, at least for the general system.[5]

If the balance of power was indeed self-balancing, Rousseau concluded, the guiding rule for foreign policy was *laissez faire.* It was a conclusion, however, that was shared by few and that had no appreciable impact on either the practical conduct of statecraft or on political theory. By the end of the eighteenth century, the meaning of the balance of power was still as indefinite as ever. Immanuel Kant simply dismissed it as a chimera.[6] Those who still tried seriously to contribute to its theory tended to go over old ground (although the proposition of self-balance was avoided) without ever agreeing.[7] Henceforth, the balance of power continued as a general rule of political common sense and as a pliable piece of rhetoric that could be bent into many different shapes. It had nothing further to do with the concept of the autonomous system.

As the constitutional monarchy that was established in Britain in 1688 proved to be an enduring practical success, it became the pride of the British and the envy of Continental Europeans. It is not clear, however, to what extent contemporary British viewers appreciated it as a self-regulating, autonomous system or at least as a deliberate attempt to devise such a system. The British constitutional government did not have the uncompromising purity of an intellectual construct because it was a living social organism that had evolved historically in the real world of practical politics. And the contemporary Englishman might have lacked detachment. Would it not be difficult for the politically engaged citizen to view the particular system within which he lived as automatic? Whatever the reasons, the vast amount of eighteenth-century English discussion of limited and balanced government never produced an entirely satisfying (at any rate, by the standards of contemporary economists) description of a self-regulating system of government.

The internal political balance of a state had traditionally been considered a special case of the balance of power. British authors expressed that view well into the eighteenth century. Jonathan Swift, for example, observed in 1701 that "it will be an eternal Rule in Politicks among every free People, that there is a Balance of Power to be carefully held by every State within itself, as well as among several States with each other,"[8] and Barrington in 1702 stated that "the constitution of England consists in a balance of parties; as the liberties of Europe do in a balance of powers."[9] Similarly, an anonymous pamphleteer (1720) described political equilibrium as "a balance of power between prince and people,"[10] and another (1743) advocated that, in order "To fix the Balance of Power in Europe, attention must be paid to the inferior Balances elsewhere."[11]

Perhaps this habit of treating the internal and external political balances alike represented a rhetorical tradition more than a considered conviction, but it had the effect of carrying the confusion about the ontological status of the balance of power—the uncertainty about whether it was a normative or a descriptive concept—over into discussions of the state's internal balances. Often the balances, or checks and balances, of the state were invoked to enjoin the audience to guard them vigilantly and to do everything required to keep them in equilibrium. Some statements by Defoe, Bolingbroke, and Blackstone are characteristic. Appealing to the conservatism of his audience, Daniel Defoe in 1698 recommended balanced government because of its roots in English tradition:

> The People obtain'd Priviledges of their own, and oblig'd the King and the Barons to accept of an Equilibrium, this we call a Parliament: And from this the Due Ballance, we have so much heard of is deduced. I need not lead my Reader to the Times and Circumstances of this, but this Due Ballance is the Foundation on which we now stand . . . and I appeal to all Men to judge if this Ballance be not a much nobler Constitution in all its Points, than the old *Gothick* [i.e., feudal] Model of Government.[12]

Bolingbroke emphasized the necessity of preserving the separation of powers; in 1730 he urged that "in a constitution like ours, the safety of the whole depends on the balance of the parts, and the balance of the parts on their mutual independence on one another."[13] And Sir William Blackstone, a generation later (1760s), made the same point:

> But the constitutional government of this island is so admirably tempered and compounded, that nothing can endanger or hurt it, but destroying the equilibrium of power between one branch and the legislature and the rest. For if ever it should happen that the independence of any one of the three should be lost, or that it should become subservient to the views of either of the other two, there would soon be an end to our constitution.[14]

There is a tone of exhortation in these quotations which they have in common with the typical balance-of-power rhetoric. But there is also a difference. The balance of power had to be achieved by the deliberate actions of distinct persons with the necessary powers. Therefore, whenever the balance of power was discussed in the literature, it tended to be with the purpose of praising its benefits and urging its preservation. The balance of government, however, was preserved by organizational arrangements within specific institutions, by something that could be called an impersonal mechanism. This mechanism was feared to be precarious and fragile. The intent of the above quotations, accordingly, was not to incite activity to maintain the balance itself but only to urge the protection of the fragile balancing mechanism.

How was this mechanism envisioned to work? An entirely satisfactory contemporary description of the mechanism is hard to find, but a selection of representative statements in chronological order may suggest the general manner and style in which the task was approached.

A popular technique for commenting on contemporary politics in a posture of objectivity was to discuss a similar political situation in the distant past. Walter Moyle, writing about 1699 on republican Rome, clearly had his mind on recent English history when he said:

> But although there was no counterpoise in the laws to balance the regal authority, yet the liberty seems to have been in a great measure supported by the mere weight of the people, whose property was the noblest root of

> liberty. . . . Each branch was a check upon the other, so that not one of them could exceed its just bounds, but was kept within the sphere in which it was circumscribed by the original frame.[15]

While expressing a confident belief in the self-regulating powers of the system, the passage does not describe a specific mechanism. Jonathan Swift (1701), however, made a serious attempt to explain the working of self-regulation in statecraft with the help of a detailed mechanical model derived from the balance analogy:

> The true Meaning of a Balance of Power, either without or within a State, is best conceived by considering what the nature of a Balance is. It supposes three Things. First, the Part which is held, together with the Hand that holds it; and then the two Scales, with whatever is weighed therein. Now consider several States in a Neighbourhood: In order to preserve Peace between these States, it is necessary they should be form'd into a Balance, whereof one or more are to be Directors, who are to divide the rest into equal Scales, and upon Occasions remove from one into the other, or else fall with their own Weight into the Lightest. So in a State within itself, the Balance must be held by a third Hand; who is to deal the remaining Power with utmost Exactness into the several Scales. Now, it is not necessary that the Power should be equally divided between these three; For the Balance may be held by the Weakest, who, by his Address and Conduct, removing from either Scale, and adding of his own, may keep the Scales duly pois'd.[16]

This account of the balancing mechanism claims to be universally applicable; it attempts to fit both the external and internal balances of power, both when managed personally by individual agents or when regulated automatically by institutional mechanisms. Actually, the account only demonstrates the limitations of the balance analogy, which fails to do justice to the actual phenomena and proves incapable of accounting for the phenomenon of self-regulation.

As the inadequacy of the balance image for the representation of self-regulating political processes became clearer, authors searched for other, more effective terminology. A term that has been encountered earlier and that was to become popular in the combination *checks and balances* occurs, for example, in James Thomson's *Liberty* (1736) with the purpose of praising, not explaining, the autonomous system:

> The full, the perfect plan
> Of Britain's matchless Constitution, mixt
> Of mutual checking and supporting powers,
> King, Lords, and Commons.[17]

The word *checking* presumably had its origins in the game of chess, but as used here it had lost its pictorial force, and its meaning was totally abstract.[18]

Bolingbroke tried to describe the self-regulating mechanism, now called a *system,* in astronomical terms:

> A king of Britain is now . . . what kings should always be, a member, but the supreme member . . . of a political body: . . . he can move no longer in another orbit from his people, and, like some superior planet, attract, repel, influence, and direct their motions by his own. He and they are parts of the same system, intimately joined and cooperating together, acting and acted upon, limiting and limited, controlling and controlled by one another; and when he ceases to stand in this relation to them, he ceases to stand in any.[19]

The mechanism, although in constant motion, maintains a form of equilibrium: when some force in it shows signs of becoming too strong, a counterforce will appear automatically to balance it. Bolingbroke was describing a *dynamic equilibrium.*

Gregory Sharpe, chaplain to the Prince of Wales, who admired "the excellency of mixed government, such as our own," was concerned "how to preserve the several parts of it in an equal poise, that one may not rise up and depress the other." He visualized the system in terms of traditional mechanical imagery: "In a compound machine, the several parts may have different powers and a different motion . . . take away one of the powers and the machine is destroyed. It is the same in mixed government."[20] In comparing the mixed government with a "compound machine," presumably some form of clockwork that would break down whenever a part of it was broken or missing, he showed a lack of either confidence in or understanding of the self-balancing capabilities of the system. This imagery made it impossible to envision a process in which any imbalance would automatically initiate corrective measures.

As discussions of constitutional government progressed, mechanical imagery lost its power. The self-balancing capability, its chief characteristic, was seen to be the result of the free play of opposing forces. In 1752 Thomas Pownall defined "mixed government" abstractly by its ability "to subsist and be carried by Parties and Oppositions: for it consists of divers and different Parties, which can only subsist by opposing and being a check upon each other."[21] Pownall was critical of contemporary abuses of the principle of balanced government, where the checks and balances were invoked to justify blatantly self-serving activities and where equilibrium degenerated into stagnation. The language used in his description of the system highlights the limitations of mechanical imagery:

> Thus it becomes the interest of the Democratic part, to be a constant Clog and Check upon the Measures of the administering Power, and to oppose themselves to every new Exertion of its influence. Here there are two differ-

> ent Parties whose interest is essentially contrary, and who can alone subsist by the Struggles of Opposition.
>
> Nay it is of the Spirit of this Policy to speak of these as set up to be a Counterbalance the one to the other, to oppose, check and impede the other. Hence a Balance of Power, and a due Regulation of this balance, is the Essence of this Constitution.[22]

David Hume, when referring to constitutional government, alternated between the slightly misleading specificity of the balance—"It is possible so to constitute a free government, as that a single person, call him doge, prince, or king, shall possess a large share of power, and shall form a proper balance or counterpoise to the other parts of the legislature"—and the convenient vagueness of the checks:

> All absolute governments must very much depend on the administration; and this is one of the great inconveniences attending to that form of government. But a republican and free government would be an obvious absurdity, if the particular checks and controuls, provided by the constitution, had really no influence, and made it not the interest, even of bad men, to act for the public good.[23]

Hume here introduced an idea that was a central element in contemporary liberal doctrine, namely, the idea of *self-interest* as the source of energy behind successful self-regulating social systems. He did not, however, spell out the details of the interdependence between self-interest and self-regulation; that was left to his friend Adam Smith.

An abundance of similar language but also a strong analytical interest in the functioning of the mechanism can be found in Blackstone's *Commentaries on the Laws of England.* Speaking of the need of "preserving the ballance of the constitution" he pointed out that "the constitutional government of this island is so admirably tempered and compounded, that nothing can endanger or hurt it, but destroying the equilibrium of power between one branch and the legislature and the rest."[24] The actual functioning of the mechanism was described in these terms:

> And herein indeed consists the true excellence of the English government, that all the parts of it form a mutual check upon each other. In the legislature, the people are a check upon the nobility, and the nobility a check upon the people; by the mutual privilege of rejecting what the other has resolved: while the king is a check upon both, which preserves the executive power from encroachments. And this very executive power is again checked, and kept within due bounds, by the two houses, . . . Thus, every branch of our civil polity supports and is supported, regulates and is regulated, by the rest.

Shifting imagery again, he then compared the three branches of government to three forces acting upon the same body whose resultant is determined by the method of the parallelogram of forces:

> Like three distinct powers in mechanics, they jointly impel the machine of government in a direction different from what either, acting by itself, would have done; but, at the same time, in a direction partaking of each, and formed out of all: a direction which constitutes the true line of the liberty and happiness of the community.[25]

There is no need to extend this survey further. Certain common characteristics in these comments by British authors on the balances of government are obvious. These attempts to describe a self-regulating government arose from the experience of actually living under constitutional government, a form of government that had proven its effectiveness and durability in practice and that was cherished by its citizenry. And these comments express an awareness that constitutional government had a unique quality, namely, the ability to maintain its own equilibrium by virtue of a certain institutionalized mechanism. All the authors of these comments, however, had difficulties describing that self-regulating mechanism and its functioning in time. A wide array of images, metaphors, and analogies was invoked to elucidate the mechanism—the balance, checks, gravitation and planetary orbits, clockwork, and parallelograms of force; they all proved inadequate, whether used by themselves or in various combinations. Since the self-regulating mechanism of government also resisted description in direct, nonmetaphorical language, it soon became conventional to refer to it simply by the shorthand formula of "a system of checks and balances" which, being entirely undescriptive, was assumed to be generally familiar.

10

Self-regulation in Economic Thought

The earliest really clear statements of the concept of self-regulation can be found in economic theory, specifically in the doctrine of economic liberalism. It might be argued that in Adam Smith's *Wealth of Nations* (1776), the book that announced economic liberalism to the world, self-regulation was the central concept. The task of the following, then, is to trace the evolution of the concept through the early history of economic liberalism up to and through *The Wealth of Nations*.

Economic liberalism as a theoretical system had evolved in Britain in the course of the eighteenth century as an alternative to the doctrine of mercantilism, which had been dominating practical economic policy in all of Europe. The roots of the new system can be found on different levels; on the level of practical politics, the central doctrine of mercantilism was fought head-on by a parliamentary faction known as the Free Traders.[1]

On the level of economic theory, economists tried to provide analytical proof that mercantilism, with its commitment to a positive trade balance, was wrong on logical grounds and that liberal economic policies would provide greater benefits for a nation governed by them. Although, on the surface, the imperative of a favorable balance of trade had the appearance of an obvious road to national wealth, its validity was questioned early on. The accumulation of gold and silver in a given state, it was suspected, did not generate a commensurate amount of well-being because the very abundance of currency set off reactions that counteracted it. The exploration of this problem led quite naturally to the concept of self-regulation.

On the level of social values, there had been a basic reorientation with regard to the individual pursuit of self-interest. What traditional ethics had once condemned as the vice of selfishness began now to be perceived as a useful quality. Even the mercantilists had appreciated the importance of the profit drive for motivating economic initiative, but they had still regarded it as a vice to be curbed by sound regulation. Liberalism, in contrast, believed that this vice could be put to work for the benefit of society; Bernard de Mandeville had argued that many of the energies upon which a healthy state depends arise from motives that are anything but noble. Gradually self-interest was accepted, if not as a virtue than at least as a source of energy that motivated initiative, enterprise, and innovation—a quality that the state should tolerate within the normal laws, without restriction and regulation. This view was also embraced by the followers of a small and short-lived but well-publicized economic movement in France, the physiocrats, who coined a name for it, *laissez faire,* by which it has come to be generally known.

For the loose slogan of laissez faire to develop into a fully articulated economic theory, an important question had yet to be answered: precisely what was it that harnessed the chaotic activities of countless self-seeking competitors on the market place into the service of the common good? The answer was to be supplied by Adam Smith in terms of the concept of self-regulation. He explained in detail how the self-balancing mechanism of the free market regulated the economy better than the most benevolent, omniscient central authority. Since this agent could not be seen anywhere, he called it the Invisible Hand.

In the evolution of the notion of self-regulation in economics, one can identify the same stages that were traversed in other areas of thought: from an unequal to an even balance; from equilibrium maintained by conscious effort to an equilibrium maintained automatically; from self-balancing capabilities in particular processes to self-regulation as a general interdisciplinary concept.

The following survey of early formulations in economic theory of the notion of self-regulation is selective. There were, for example, mercantilist writers both on the Continent and in Britain who perceived limitations on the effective use of the principle of the positive trade balance. Such observations occasionally have been interpreted as anticipations of the notion of self-regulation.[2] Unless they make some effort to give a causal explanation of self-regulation, or unless they show some awareness of the wider significance of the principle, they will not be considered here. Some economic writers touched upon self-regulation only incidentally in the context of some particular economic problem whereas others seemed genuinely fascinated with it. Authors of the latter kind would naturally tend to give clearer accounts of the concept and will therefore be favored here.

We begin our sampling with a little-known economic writer who is remarkable for having recognized self-regulation as a concept of independent significance at an early point. Isaac Gervaise was a London businessman of Huguenot extraction who in 1720 published a book called *The System or Theory of the Trade of the World.*[3] He invoked self-regulation repeatedly, not only in connection with standard economic problems but also to interpret social phenomena that might just as well have submitted to more conventional explanations.

A natural occasion for introducing self-regulation was the discussion of the familiar mercantilist doctrine of the positive trade balance, that is, of the belief that the way to increase a country's wealth was to maximize its holdings of gold and silver by stimulating export and restricting import. Gervaise argued that the holdings of currency in a country, in proportion to its population, oscillated around a natural level; there was an automatic mechanism that, after temporary deviations, restored those holdings of currency to their natural level.

To follow Gervaise's exposition, it will help to know that by "the grand denominator of the world" he meant currency, and by "the poor" he meant the working population; he also considered a country's rate of production to be in direct proportion to the size of its working population. The pertinent passage deserves to be repeated in full:

> When a Nation has attracted a greater Proportion of the grand Denominator of the World, than its proper share; and the Cause of that Attraction ceases, that Nation cannot retain the Overplus of its proper Proportion of the grand Denominator, because in that case, the Proportion of Poor and Rich of that Nation is broken; that is to say, the number of Rich is too great, in proportion to the Poor, so as that Nation cannot furnish unto the World that share of Labour which is proportion'd to that part of the grand Denominator it possesses: in which case all the Labour of the Poor will not ballance the Expence of the Rich. So that there enters in that Nation, more Labour than goes out of it, to ballance its want of Poor: And as the End of Trade is the attracting Gold and Silver, all that difference of Labour is paid in Gold and Silver, until the Denominator be lessen'd, in proportion to other Nations; which also, and at the same time, proportions the number of Poor to that of Rich.
>
> Thus as Labour draws the Denominator of the World, also the Denominator draws Labour from the World; so that if the particular Denominator of any Nation, be greater than its just Proportion, it will draw from the other Nations a Portion of Labour, proportion'd to its Excess; and if its Denominator be less than its just Proportion, it will draw a Portion of Gold and Silver, proportion'd to what it wants of its just Proportion.[4]

Translation of the passage into block diagram notation (see figure 10–1) will reveal Gervaise's self-regulating mechanism as a closed feedback loop. Gervaise identified two concentric feedback loops: the increasing wealth of a country not only reduces the proportion of the population that is willing to work but also increases its domestic consumption and thus decreases its exports. Gervaise concluded that

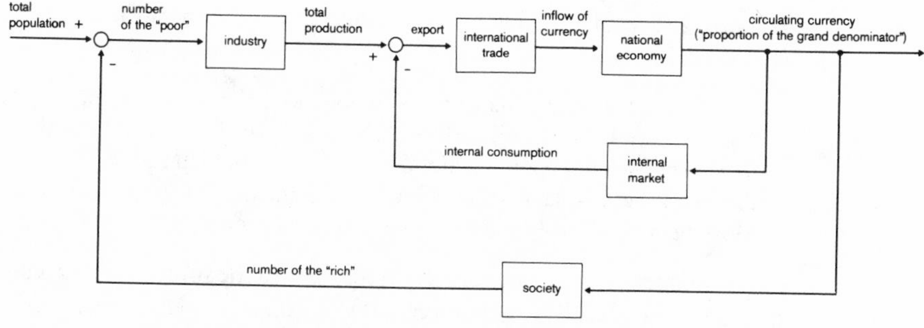

Figure 10–1. Isaac Gervaise's "natural ballance of trade" (1720).

"Trade causes a Vibration, or continual Ebbing and Flowing; which may be called the natural Ballance of Trade" and

> That if Trade was not curbed by Laws, or disturbed by those Accidents that happen in long Wars, etc. which break the natural Proportion, either of People, or of private Denominators; Time would bring all trading Nations of the World into that Equilibrium, which is proportioned, and belongs to the number of their Inhabitants.[5]

A liberal, antimercantilist economic philosophy can be detected in these arguments, but it is not expressed with force. Gervaise only presented the results of an analysis but did not draw theoretical or practical conclusions.

Notions of self-regulation also surfaced on other occasions in Gervaise's book, even when not provoked by the ideological conflict between mercantilism and liberalism. Discussing, for example, the motivations that impelled people in their business activities, Gervaise perceived an equilibrium of two opposite appetites, the desire for wealth and the love of leisure. Each counteracted the other in such a way so as never to permit either of them to predominate, thus keeping both desires permanently sharp and active:

> As Gold and Silver not only express the Value of things; but also carry with them a Right, or Demand at will, on all things necessary: all Men have, one with the other, an equal desire to draw them to themselves; which can be done, but by Labour only: And as Man naturally loves his Ease, the Possession of a part of them lessens his Desires, and causes him to labour less; which gives him that hath little or no Possession (and consequently preserves his Desire intire) an opportunity by his Labour to slip into his place.[6]

In analyzing this formulation, one must bear in mind that Gervaise was trying to express something that was both abstract and novel.

Perhaps aware of the difficulty, he offered a mechanical analogy for clarification:

> This Desire [of wealth] may be look'd upon as the great Spring that forces Movement or Labour; and the love of Ease, as the small Spring or Pendulum, that keeps Men in a continual Equilibral Vibration of Rich and Poor; so that the one always ballances the other, in such manner, as keeps Labour or Movement continually going, in a certain equal proportion.[7]

The imagery is derived from the clock, where the force of the driving spring is regulated by an escapement with either a balance spring or a pendulum. The analogy is hard to visualize because it does not fit. Actually, Gervaise was describing a genuine feedback system (see figure 10–2); the continuous oscillation of such a system around an equilibrium position, which he described perceptively, is a characteristic feature of such systems.

Gervaise's theory of the self-regulating interaction of man's desire of wealth and his love of ease was of no relevance to economic theory and of uncertain merit as a contribution to psychology. The passage is interesting, however, as a statement of the concept of self-regulation and significant as one that was suggested not from practical experience but that was inspired, it seems, by an a priori predisposition toward this concept.

An economist contemporary to Gervaise whose views were similarly advanced but who lacked Gervaise's fascination with self-regulation was Richard Cantillon (1680?–1743), an Irish banker living in Paris, who wrote, around 1730 and in English, a treatise on economics which was published in French translation only in 1755.[8] An understanding of the self-equilibrating nature of various economic phenomena is indicated there repeatedly, but such formulations are vague and are made without interest and engagement and without attention to the actual mechanism that accomplishes the self-regulation. For example, in a chapter entitled "The Number of Labourers, Handicraftsmen and Others, Who Work in a State is Naturally Proportioned to the Demand for

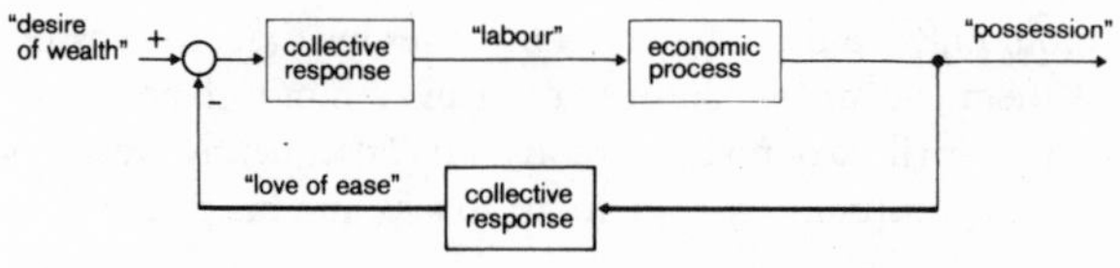

Figure 10–2. Isaac Gervaise's theory of the equilibrium between the desire for wealth and the love of ease.

Them," he had no more to say than that "it is easy to conceive that the labourers, handicraftsmen and others who gain their living by work, must proportion themselves in number to the employment and demand for them in market towns and cities."[9] When dealing with the self-regulation of market prices through the action of supply and demand, he only observed: "The butcher keeps up his price according to the number of buyers he sees; the buyers, on their side, offer less according as they think the butcher will have less sale."[10] His description of the mechanism that regulates the international distribution of currency was inferior to Gervaise's. Cantillon's best description of self-regulation is a short passage explaining the mechanism that regulates the size of a given trade, in this case that of the hatters:

> If there are too many hatters in a city or in a street for the number of people who buy hats there, some who are least patronized must go bankrupt; whereas if there are too few, it will be a profitable enterprise, which will encourage some new hatters to open shops there; and it is in this way that entrepreneurs of all kinds proportion themselves to the risk in a state.[11]

As a description of a somewhat trivial self-regulating system, this is adequate enough; it leaves no doubt that Cantillon was capable of appreciating the existence and comprehending the functioning of such systems. The insignificance of the examples and the lack of emphasis in his overall treatment suggest, however, that Cantillon had no interest in self-regulation for its own sake and was not inclined to assign to it a role of wider significance.

Quite different is the role that the concept of self-regulation played in David Hume's strongly antimercantilist essay, "Of the Balance of Trade," written about 1750.[12] "Balance" to Hume meant equilibrium. The essay's purpose was to prove that the balance of trade maintained itself in equilibrium automatically. Too many, he stated in the beginning, did not yet know this. There was not only the "gross and palpable" error of "nations ignorant of the nature of commerce, to prohibit the exportation of commodities, and to preserve among themselves whatever they think valuable and useful," a misconception that resulted in stifling the country's native productivity, but also "there still prevails, even in nations well acquainted with commerce, a strong jealousy with regard to the balance of trade, and a fear, that all their gold and silver may be leaving them."[13] To expose this "groundless apprehension" and to show that the supply of currency in a state will never run out as long as its people and industry continue to prosper, he examined the response of the balanced system to a sudden disturbance:

> Suppose four-fifth of all the money in GREAT BRITAIN to be annihilated in one night, and the nation reduced to the same condition, with regard to specie, as

> in the reigns of the HARRYS and EDWARDS, what would be the consequence? Must not the price of all labour and commodities sink in proportion, and everything be sold as cheap as they were in those ages? What nation could then dispute with us in any foreign market, or pretend to navigate or to sell manufactures at the same price, which to us would afford sufficient profit? In how little time, therefore, must this bring back the money which we had lost, and raise us to the level of all the neighbouring nations? Where, after we have arrived, we immediately lose the advantage of the cheapness of labour and commodities; and the farther flowing in of money is stopped by our fulness and repletion.

The system, in other words, by its own strength, would soon restore its equilibrium. The same would happen after a disturbance in the opposite direction:

> Again, suppose, that all the money of GREAT BRITAIN were multiplied fivefold in a night, must not the contrary effect follow? Must not all labour and commodities rise to such an exorbitant height, that no neighbouring nations could afford to buy from us; while their commodities, on the other hand, became comparatively so cheap, that, in spite of all the laws which could be formed, they would be run in upon us, and our money flow out; till we fall to a level with foreigners, and lose that great superiority of riches, which had laid us under such disadvantages?[14]

This is a clear and complete description of the mechanism that automatically balances the supply of money in neighbouring countries, a mechanism that replaced or at least clarified the ambiguous concept of the balance of trade (see figure 10–3).

As a physical analogy for the functioning of this mechanism, for which the analogy of the balance was clearly no longer serviceable, Hume suggested an ingenious substitute:

> All water, wherever it communicates, remains always at the same level. . . . Were it to be raised in any one place, the superior gravity of that part not being balanced, must depress it, till it meet a counterpoise; and . . . the same cause, which redresses the inequality when it happens, must for ever prevent it, without some violent external operation.[15]

The natural mechanism that maintains the horizontal level in a body of water is, of course, the simplest self-regulating system conceivable. The mechanism cannot work successfully, however, for bodies of water that are separate, by nature or artifice.

> But as any body of water may be raised above the level of the surrounding element, if the former has no communication with the latter; so in money, if the communication be cut off, by any material or physical impediment, (for all laws alone are ineffectual) there may, in such a case, be a very great inequality of money. . . . The only expedient by which we can raise money

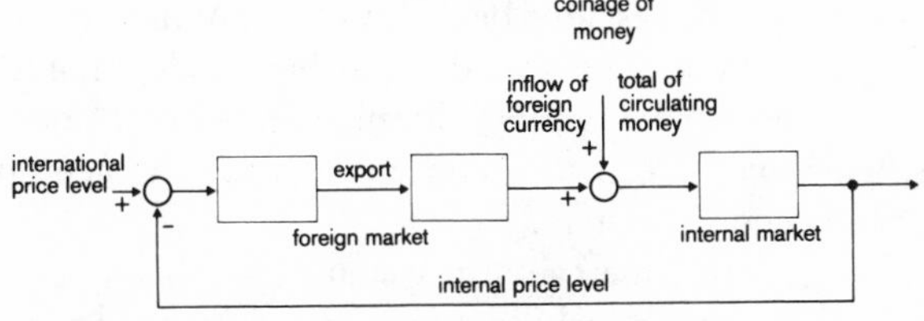

Figure 10–3. David Hume's self-regulating mechanism for international monetary distribution.

> above it, is a practice which we should all exclaim against as destructive, namely, the gathering of large sums into a public treasure, locking them up, and absolutely preventing their circulation. The fluid, not communicating with the neighboring element, may, by such an artifice, be raised to what height we please.[16]

The consequences, however, would be unwelcome:

> A weak state, with an enormous treasure, will soon become prey to some of its poorer, but more powerful neighbours. A great state would dissipate its wealth in dangerous and ill-concerted projects; and probably destroy, with it, what is much more valuable, the industry, morals, and numbers of its people. The fluid, in this case, raised to too great a height, bursts and destroys the vessel that contains it; and mixing itself with the surrounding element, soon falls to its proper level.[17]

These were the essentials of Hume's theory of the mechanism of international trade. The concept of self-regulation is applied here with a clarity, accuracy, and depth of understanding not achieved before. The concept is also unmistakably linked with the advocacy of a liberal policy. It is not possible, however, to claim Hume as someone who delighted in the concept of self-regulation for its own sake; other uses of the concept by Hume are not known.

The grand conclusion of the interdependent, almost symbiotic evolution of the concepts of self-regulation and the liberal system of economics was reached in Adam Smith's classic *Inquiry into the Nature and Causes of the Wealth of Nations,* published in 1776.[18] This book consists with equal emphasis of a polemic against the economic policies of mercantilism, of a plea on behalf of economic liberalism (Adam Smith called it "the System of Natural Liberty"), and of a blueprint for the introduction of that system into practical government. The concept of self-regulation is the unacknowledged heart of this system. *The Wealth of Nations* has occasionally been characterized as a compilation containing little that has not been presented before by others. Whatever the merits of that charge, the book is novel and original in the depth of its understanding of the functioning and significance of self-regulation, and in the consistency with which this concept is made the central element of an entire economic system.

Adam Smith charged the mercantilist system with being economically unsound, that is, with inability to achieve its avowed goals and with violation of a basic human right—freedom. He argued that these two failures were interdependent: repression was bad economics, and the natural path to a maximum of prosperity was liberalism. Mercantilism was "in its nature and essence a system of restraint and regulation." Its crucial error was its preoccupation with the balance of trade:

> The encouragement of exportation, and the discouragement of importation, are the two great engines by which the mercantile system proposes to enrich every country. . . . Its ultimate object . . . is always the same, to enrich the country by an advantageous balance of trade.[19]

Concerning the latter, he observed that

> nothing, however, can be more absurd than this whole doctrine of the balance of trade, upon which, not only these restraints, but almost all the other regulations of commerce are founded. When two places trade with one another, this doctrine supposes that, if the balance be even, neither of them either loses or gains; but if it leans in any degree to one side, that one of them loses, and the other gains in proportion to its declension from the exact equilibrium. Both suppositions are false. A trade which is forced by means of bounties and monopolies, may be, and commonly is disadvantageous to the country in whose favour it is meant to be established, as I shall endeavour to shew hereafter. But that trade which, without force or constraint, is naturally and regularly carried on between any two places, is always advantageous, though not always equally so, to both.[20]

If mercantilism was so flawed, who benefited from it? "It cannot be very difficult to determine who have been the contrivers of this whole mercantile system; not the consumers, we may believe, whose interest has been entirely neglected; but the producers, whose interest has so carefully been attended to." This is repeated elsewhere: "But in the mercantile system, the interest of the consumer is almost constantly sacrificed to that of the producer." The ultimate instrument for protecting a given producer was the monopoly: "Monopoly of one kind or another, indeed, seems to be the sole engine of the mercantile system."[21] Mercantilism, then, was not only logically faulty and morally untenable in a general way; it was also quite specifically vicious by being a cold-blooded attempt by a small minority to despoil society at large.

The alternative system that Smith proposed was completely subordinated to the principle of freedom. Commercial freedom was the precondition for the functioning of his system and the source of the unequaled benefits that were to flow from it. This was reiterated many times, notably in some very specific arguments. The "natural price" (i.e., cost) of goods, for example, would come down to the lowest possible point "where there is perfect liberty." Conversely, the "market price [should it ever be below cost] . . . would soon rise to the natural price. This at least would be the case where there was perfect liberty."[22]

Similarly, in the absence of all restraint, the total of the comparative advantages and disadvantages of all occupations would tend toward equality: "This at least would be the case in a society where things were left to follow their natural course, where there was perfect liberty, and where every man was perfectly free both to chuse what occupation he thought proper, and to change it as often as he thought proper." The beneficial effects of freedom were also forecast on a larger scale for society in general: "The establishment of perfect justice, of perfect liberty, and of perfect equality, is the very simple secret which most effectually secures the highest degree of prosperity to all three classes."[23] With regard to agricultural nations lacking industries, Smith stated:

> This perfect freedom of trade would even be the most effectual expedient for supplying them, in due time, with all the artificers, manufacturers and merchants, whom they wanted at home, and for filling up in the properest and most advantageous manner that very important void which they felt there.[24]

Commercial freedom meant absence of regulation. Even without planning and supervision by an attentive government, a nation with commercial freedom would always have the commodities it needed without exhausting its supply of currency. The balance of trade, in other words, would always balance itself:

> We trust with perfect security that the freedom of trade, without any attention of government, will always supply us with the wine which we have occasion for: and we may trust with equal security that it will always supply us with all the gold and silver which we can afford to purchase or employ.[25]

Pleas of this kind for freedom of trade abound throughout *The Wealth of Nations;* several have been repeated above, at the risk of monotony, to convey the emphasis with which Adam Smith treated the subject.

Pleas for economic freedom, however, are not an economic system. After a discussion of the economic systems of the mercantilists and physiocrats, Smith presented his own "system of natural liberty" in programmatic and comprehensive manner:

> All systems either of preference or of restraint, therefore, being thus completely taken away, the obvious and simple system of natural liberty establishes itself of its own accord. Every man, as long as he does not violate the laws of justice, is left perfectly free to pursue his own interest his own way, and to bring both his industry and capital into competition with those of any other man, or order of men. The sovereign is completely discharged from a duty, in the attempting to perform which he must always be exposed to innumerable delusions, and for the proper performance of which no human wisdom or knowledge could ever be sufficient; the duty of superintending the industry of private people, and of directing it towards the employments most suitable to the interest of the society. According to the system of natural liberty, the sovereign has only three duties to attend to; three duties of great importance, indeed, but plain and intelligible to common understandings: first, the duty of protecting the society from the violence and invasion of other independent societies; secondly, the duty of protecting, as far as possible, every member of the society from the injustice or oppression of every other member of it, or the duty of establishing an exact administration of justice; and, thirdly, the duty of erecting and maintaining certain public works and certain public institutions, which it can never be for the interest of any individual, or small number of individuals, to erect and maintain; because the profit could never repay the expence to any individual or small number of individuals, though it may frequently do much more than repay it to a great society.[26]

This paragraph sums up Adam Smith's system of economic liberalism. It remains to be explained how this system was to function and how it was to deliver the amazing benefits that Smith had promised for it.

As the motor of the system, as its source of initiative and drive, the government, especially the sovereign, was now replaced by an agent that was not a person or an institution but merely a human characteristic, namely, man's overriding preoccupation with his own self-interest:

> Man has almost constant occasion for the help of his brethren, and it is vain to expect it from their benevolence only. He will be more likely to prevail if he can interest their self-love in his favour, and shew them that it is for their own advantage to do for him what he requires of them. . . . It is not from the benevolence of the butcher, the brewer or the baker, that we expect our dinner, but from their regard to their own self-interest.[27]

Self-interest was the cause of a phenomenon indispensable to economic life, competition: "Good management . . . can never be universally established but in consequence of that free and universal competition which forces everybody to have recourse to it for the sake of self-defence." And self-interest was the fuel that fired people's imagination and inventiveness and powered their enterprise and initiative:

> Every individual is continually exerting himself to find out the most advantageous employment for whatever capital he can command. It is his own advantage, indeed, and not that of society, which he has in view. But the study of his own advantage naturally, or rather necessarily leads him to prefer that employment which is most advantageous to the society.[28]

It was plausible enough that self-interest would incite competition and stimulate industry, initiative, and ingenuity. But how would its unrestrained pursuit lead to anything but chaos? Adam Smith answered the question with a concept that had the appearance of mystery; in the economic system of natural liberty, an Invisible Hand somehow took charge of coordinating the self-serving activities of individuals for the benefit of the whole community:

> Every individual necessarily labours to render the annual revenue of the society as great as he can. He generally, indeed, neither intends to promote the public interest, nor knows how much he is promoting it. By preferring the support of domestic to that of foreign industry, he intends only his own security; and by directing that industry in such a manner as its produce may be of the greatest value, he intends only his own gain, and he is in this, as in many other cases, led by an *invisible hand* [emphasis added] to promote an end which was not part of his intention. Nor it is always the worse for the society that it was no part of it. By pursuing his own interest he frequently promotes that of the society more effectually than when he really intends to promote it. I have never known much good done by those who affected to trade for the public good.[29]

Smith's Invisible Hand could not be identified with a specific person, institution, or program or with a definite bureaucratic mechanism. It was an abstract power immanent in the system. The capability to employ opposing, uncooperative forces to establish and maintain equilibrium is a characteristic of self-regulating or—in cybernetic jargon—feedback systems. The Invisible Hand was nothing but the quality of self-regulation.

Smith did not define and describe the functioning of the Invisible Hand in a generalized terminology. He had no generic names for the concept of self-regulation and for the economic processes characterized by it. He simply described them in the course of his narrative whenever the occasion arose. Those descriptions, however, are frequent, and they are splendidly clear. In the very beginning of *The Wealth of Nations* are three detailed expositions of specific self-regulating processes; cursory allusions to such processes are frequent throughout the remainder of the book.

The most explicit of these accounts deals with the mechanism of the market, be it the market of commodities or of labor. First, Smith formulated in complete generality what is now popularly called "the law of supply and demand."[30] The Austrian economist J. A. Schumpeter, who was no admirer of the Scottish philosopher, has praised this treatment as "by far the best piece of economic theory turned out by A. Smith."[31] The analysis of this system is presented in language so clear and is conceptualized so generally that it can be translated into the notation of modern systems theory without any need for addition or modification.

The system comprises two processes, market and industry, and four variables, the demand and supply of a given commodity, its market price, and its natural price (i.e., actual cost). Two of these variables, demand and natural price, act upon the system independently from the outside; the other two, supply and market price, are internal variables dependent upon the characteristics of the system. Of special interest is the regulated variable, the supply: "The quantity of every commodity brought to market naturally suits itself to the effectual demand."[32] Given a free economy, it is postulated, the supply will always adjust itself so as to be proportional to the demand.

The system forms a closed feedback loop. Smith defined this loop by describing its elements in proper sequence. He began at the point where the independent variable, demand, and the dependent feedback variable, supply, are compared (forming a point of subtraction or "summing junction" that every feedback system must have). In real life, the place where this comparison is made is the market:

> The market price of every particular commodity is regulated by the proportion between the quantity which is actually brought to market, and the demand of those who are willing to pay the natural price of the commodity, or the whole value of the rent, labour, and profit, which must be paid in order to bring it thither.[33]

Mathematically then, the market price is a function of the difference between demand and supply (see figure 10–4). This bald formula, to be sure, does not do justice to the extremely complex social processes for which it stands. To substantiate the formula, Smith described the workings of the market and the formation of prices at great length for the three possible cases: for the supply being smaller,[34] greater,[35] or equal[36] to the demand. The prices that can be obtained on the market are being constantly watched by the producers, for the market price minus the

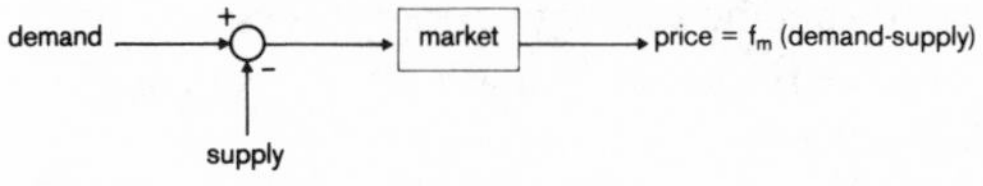

Figure 10–4. Adam Smith's general theory of supply and demand—1: price as a function of demand minus supply.

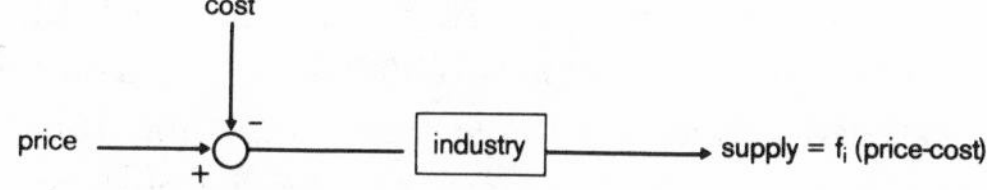

Figure 10–5. Adam Smith's general theory—2: supply as a function of price minus cost.

cost of production (called by Adam Smith the "natural price") equals the profit available to producer and distributor (for simplicity, here called the "industry"). The lower the profit in a given product, the less the industry will produce of it and supply to the market;[37] the higher the profit, the more it will supply.[38] The relationship between price, cost, and supply can again be expressed mathematically (see figure 10–5).

Supply and demand, as established earlier, control the market price. Now it becomes clear that the market price, in turn, controls the supply. Thus the whole process becomes circular, with supply acting as the feedback variable that closes the loop (see figure 10–6). The system is self-regulating because an *effect* automatically counteracts its own *cause.* A short supply, for example, via high prices and profits, will cause the supply to increase:

> The whole quantity of industry annually employed in order to bring any commodity to the market, naturally suits itself in this manner to the effectual demand. It naturally aims at bringing always that precise quantity thither which may be sufficient to supply, and no more than supply, that demand.[39]

So far we have the description of the basic functioning of the system. Added were some remarks on the transient behavior of the system in response to disturbances. The equilibrium of the system is a dynamic one; its variables must not be expected to remain rigidly constant. In the event of disturbances, there would be certain "occasional and temporary fluctuations in the market price of any commodity,"[40] but they would oscillate pendulum-like around an average value:

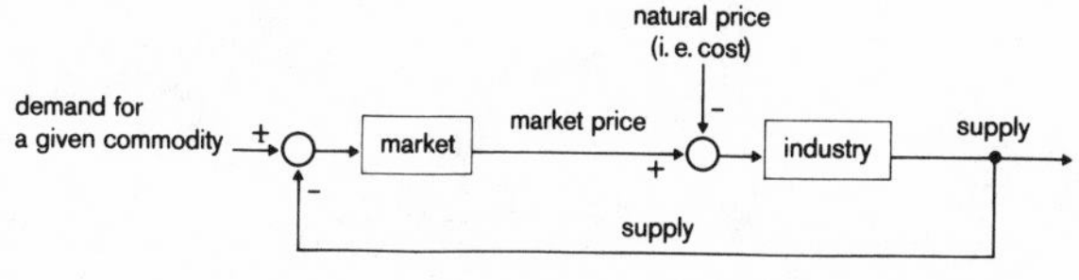

Figure 10–6. Adam Smith's general theory—3: the complete system.

> The natural price, therefore, is, as it were, the central price, to which the prices of all commodities are continually gravitating. Different accidents may sometimes keep them suspended a good deal above it, and sometimes force them down even somewhat below it. But whatever may be the obstacles which hinder them from settling in this center of repose and continuance, they are constantly tending towards it.[41]

This, then, was a demonstration of the workings of the Invisible Hand. Its capabilities were impressive; it not only automatically adjusted the supply of all possible goods and services to the demand but also took care that these were offered at the most reasonable price. The description of the generalized supply-and-demand mechanism was Adam Smith's first and most extensive exposition of self-regulation. Later in *The Wealth of Nations,* such systems were treated more and more briefly; two of them deserve detailed notice here.

A system that regulated the supply of labor in proportion to demand was closely modeled on the general supply-and-demand system.[42] There were the same variables: the demand and supply of labor (i.e., the size of the working population), the wage level (corresponding to market price), and the subsistence level (corresponding to natural price). The interaction of these variables was analyzed in two processes, the population dynamics of the working class and the labor market (see figure 10–7).

Smith began with a long account of how the size of the working population proportioned itself directly to its income level. If income should fall below the subsistence level, the population would decline, through a decrease of marriages and an increase in infant mortality:

> Every species of animals naturally multiplies in proportion to the means of their subsistence, and no species can ever multiply beyond it. But in civilized society it is only among the inferior ranks of people that the scantiness of subsistence can set limits to the further multiplication of the human species; and it can do so in no other way than by destroying a great part of the children which their fruitful marriages produce. . . .
>
> The liberal reward of labour, by enabling them to provide better for their children, and consequently to bring up a greater number, naturally tends to widen and extend those limits.[43]

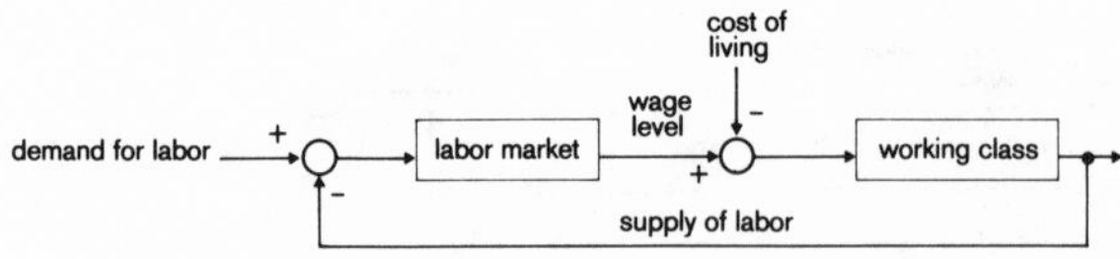

Figure 10–7. Adam Smith's theory of the supply and demand of labor.

Mathematically then, the size of the working population is directly proportional to the difference between the wage level and the subsistence level.[44]

The wage level, however, is controlled by the labor market, in proportion to the currently prevailing supply and demand:

> If the demand is continually increasing, the reward of labour must necessarily encourage in such a manner the marriage and multiplication of labourers, as may enable them to supply that continually increasing population. If the reward should at any time be less than what was requisite for this purpose, the deficiency of hands would soon raise it; and if it should at any time be more, their excessive multiplication would soon lower it to this necessary rate. The market would be so much understocked with labour in the one case, and so much overstocked in the other, as would soon force back its price to that proper rate which the circumstances of the society required.[45]

The two processes, labor population and labor market, are linked, like the generalized supply-and-demand system, in circular manner, with each of the two automatically controlling the other: "It is in this manner that the demand for men, like that for any other commodity, necessarily regulates the production of men; quickens it when it goes on too slowly, and stops it when it advances too fast."[46]

A few pages later, Smith illustrated the working of the Invisible Hand through yet another closed-loop system, this time the system that is claimed to regulate overall social justice.[47] The presence of this system assured that the sum total of the rewards and the contributions of a worker (broadly defined to include risk, costliness of training, and so forth) would be the same for all occupations in a given community—in the absence of artificial restrictions:

> The whole of the advantages and disadvantages of the different employments of labour and stock must, in the same neighbourhood, be either perfectly equal or continually tending to equality. If in the same neighbourhood, there was any employment evidently either more or less advantageous than the rest, so many people would crowd into it in the one case, and so many would desert it in the other, that its advantages would soon return to the level of other employments. This at least would be the case in a society where things are left to follow their natural course, where there was perfect liberty, and where every man was perfectly free both to chuse what occupation he thought proper, and to change it as often as he thought proper. Every man's interest would prompt him to seek the advantageous, and to shun the disadvantageous employment.[48]

This account again suggests two processes: 1. the available workers comparing all the various possible employments and rushing into the most advantageous one (and vice versa), and 2. the diminishing of

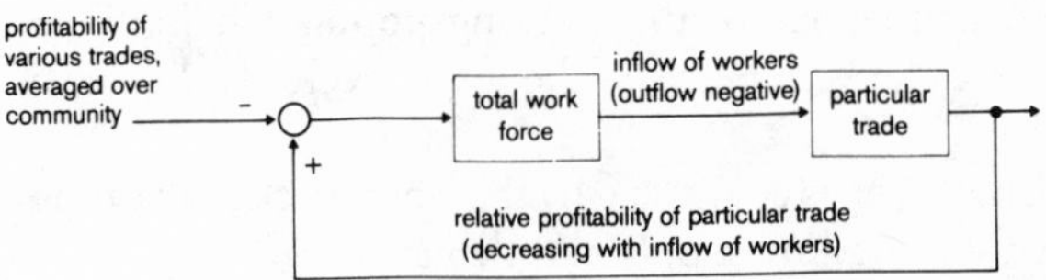

Figure 10–8. Adam Smith's theory of the self-balancing nature of the rewards of all occupations.

rewards in any particular employment in proportion to the number of workers competing in it (and vice versa). Since the result of the second process counteracts the cause of the first, the system again forms a negative feedback loop (see figure 10–8).

With these three examples, Adam Smith seems to have considered that the concept of self-regulation was explained clearly enough. Thereafter (The three examples were all presented in the first 100 out of over 900 pages!) the concept was treated as known and was used implicitly, without detailed explanation, many times in applications of increasing subtlety.

11

Self-regulation and the Liberal Conception of Order

The evolution of the concept of self-regulation or feedback has now been sketched up to the point of its full formulation. This was done by discussing relevant passages from individual authors. Quotations from a small number of authors, no matter how conceptually clear, cannot convey, however, any idea of the degree of popularity and the breadth of acceptance of the concept. The concept of self-regulation was at the center of a new liberal conception of order which was antithetical to the older authoritarian one. In this chapter, I will attempt to relate this rather technical philosophical concept of self-regulation to the broader background of social attitudes and mentalities.

Before the rise of liberalism in Great Britain, liberty and order tended to be viewed as mutually exclusive opposites, and the notion of a liberal order would have been dismissed as a contradiction in terms. A prosperous, highly civilized commercial nation, no matter how sincerely commited to liberty, could not function without some form of order. In the seventeenth century, when Britain began to base its political and social life on liberal principles, it faced the challenge of devising a conception of order that did not compromise its ideal of liberty. The task was carried out in a broad and gradual social and intellectual process. On the level of its practical everyday activities, the population in time acquired the values, attitudes, and forms of behavior that life in a free society both nurtures and demands—a process too complex to be analyzed within the constraints of this study. On the intellectual level, as documented in various philosophical discussions, an array of ideas emerged which began to outline a form of order compatible with the principles of liberty. Some of these ideas are reviewed here.

If liberty means autonomy of the individual in the absence of any compulsion exercised by a superior authority, then the chief problem of a liberal society must be the inevitability of conflicts among its members, conflicts arising from collisions between freely moving individuals, be it in commerce, politics, or religion. The authoritarian conception of order knew only one solution for this problem—regulation and regimentation.

The liberal movement, motivated as it was by the commercial enterprise of a rapidly developing industrial society, not only rejected regimentation but also, from early on, declared the pursuit of self-interest a fundamental right of the individual, even before there was a clear idea of how such self-interest was to be kept in bounds. Latitudinarian churchmen of the late seventeenth century quite normally sanctioned the pursuit of self-interest as an acceptable motive of Christian behavior.[1]

Once self-interest was recognized as a legitimate human motive, it became possible to view the problem of conflict between individuals or groups dispassionately. A variety of arguments was advanced with the purpose of defusing conflict, recommending compromise, urging tolerance, and suggesting that conflict was a blessing in disguise. In arguments of this kind, the image of the balance and the value of equilibrium once again played important roles.

If a balance poised in equilibrium symbolized a desirable state, then the weights on both sides, although opposing and counteracting each other, were necessary and valuable. Applied to social conflict, this consideration suggested a new way of regarding one's opponent. If it was desirable to preserve the equilibrium, either within one nation or among several, by arranging and rearranging alliances so that two opposing sides would always be in balance, one had to regard the competing forces not as friends or foes but as indispensable participants in a necessary process, whose behavior, even when hostile, was dictated by natural drives and objective constraints but not by malice and inherent evil.

Adversary relationships, it followed, were necessary and possibly constructive. The classical model of such a relationship, that between prosecution and defense at courts of law, was in the mind of Richard Hooker, an eminent Elizabethan churchman, when he observed, on the subject of human behavior, that the

> Appetite is the Will's solicitor, and the Will is the Appetite's controller; what we covet according to the one by the other we often reject; neither is any other desire termed properly Will, but that where Reason and Understanding, or the show of Reason, prescribeth the thing desired.[2]

Hooker was describing the human problem of ethical conduct not as a conflict between good and evil but as an interaction between two opposing psychological forces that were not subject to moral evaluation.

The new view on conflict was reflected in the practical politics of constitutional government; new standards emerged for political conduct. The new politician treated political conflict as a game with its own rules and his opponents as legitimate participants whose behavior was dictated by their assigned roles. His loyalty was not only to his immediate faction but also to the peace, prosperity, and equilibrium of the whole society. He accepted the existing forces as real but knew that they were apt to change. He considered emotions irrelevant and appreciated the benefits of compromise. To the old order, this new politician was as abominable as he was incomprehensible.

George Savile, First Marquess of Halifax, who by a series of adroit and complex political maneuvers had done more than anyone else to bring about the bloodless Glorious Revolution of 1688 and the permanent establishment of constitutional government in Britain and who in the process had repeatedly given the appearance of changing sides, found himself stigmatized as a "trimmer." He replied by writing *The Character of a Trimmer* in 1688. Far from being an apology, the book was a compendium on political behavior under the new liberal conception of order. The trimmer, traditionally taken to be a spineless opportunist without loyalty or principle, was now portrayed by Halifax as a man of moderation, dedicated to the defusing of conflict, to emphasizing the common ground, to negotiating compromises, and to balancing opposites:

> Our Climate is a *Trimmer,* between that part of the World where men are Roasted, and the other where they are Frozen; . . . our Church is a *Trimmer* between the Phrenzy of Platonick Visions, and the Lethargick Ignorance of Popish Dreams; . . . our Laws are *Trimmers,* between the Excess of unbounded Power, and the Extravagance of Liberty not enough restrained; . . . true Virtue hath ever been thought a *Trimmer,* and to have its dwelling in the middle between the two Extreams; . . . even God Almighty himself is divided between his two great Attributes, his Mercy and his Justice.
>
> In such company, our *Trimmer* is not asham'd of his Name, and willingly leaveth to the bold Champions of either Extream, the Honour of contending with no less Adversaries, than Nature, Religion, Liberty, Prudence, Humanity and Common Sense.[3]

The equilibrium of the whole, then, was the trimmer's chief goal and balancing, that is, compromise, his principal technique. In describing the latter, Halifax employed a new metaphor of equilibrium that was drawn from the original meaning of the term *trimmer:*

> This innocent word *Trimmer* signifieth no more than this, That if Men are together in a Boat, and one part of the Company would weigh it down on one side, another would make it lean as much to the contrary; it happeneth there is a third Opinion of those, who conceive it would do as well, if the Boat went

> even, without endangering the Passengers; now 'tis hard to imagin by what figure in Language, or by what Rule in Sense this cometh to be a fault, and it is much more a wonder it should be thought a Heresy.[4]

The trimmer's behavior described here was commendable, not only in politics but in all situations of life. If Halifax's arguments appear unprovocative today, it is because his wisdom has long since become conventional.

As imagery of balance and equilibrium was employed to plead for moderation between political opponents, it also served as a point of toleration among ideological adversaries. Typical is an argument in favor of religious freedom, made in 1668 by Sir Charles Wolseley, a Cromwellian partisan living in retirement: "So many divided Interests and Parties in Religion, are much less dangerous than any, and may be prudently managed to ballance each other, and to become generally more safe, and useful to a State, than any united party or interest whatever."[5] Although his advice sounded superficially like the cynical "divide et impera" of Imperial Rome, he was actually repeating the classical liberal formula: leave things alone and they will manage themselves.

William Penn constructed a similar argument, albeit from somewhat different premises. To him, too, equilibrium was a fundamental value, but he viewed it as something that had to be maintained through deliberate action. After enumerating various great historical leaders who had succeeded precisely because of their tolerance of religious or ethnic diversity among their followers, he presented this physical illustration:

> In nature we see all heat consumes, all cold kills: that three degrees of cold to two of heat, allay the heat, but introduce a contrary quality, and overcool by a degree: but two degrees of cold to two of heat, make a poize in elements, and a balance in nature. And in those families where the evenest hands is carried, the work is best done, and the master is most reverenced.

He went on "to another benefit which accrues to the monarchy by a toleration, and that is a balance at home" and concluded by urging moderation upon any government that is forced to take action against excessive disobedience, that is, "not to destroy the balance, lest it should afterwards want the means of overpoizing faction."[6]

Penn seems to have had some intuition of the potential for self-regulation of social systems and to have perceived vaguely that these capabilities would function best in the absence of interfering authorities. But he did not carry this reasoning through to its logical conclusion, which would have prohibited government interference altogether; he only counseled that such intervention, when inevitable, should be moderate and delicate.

Concerning the question of how to reconcile the pursuit of individual self-interest with the good of the whole community, Penn had two answers. The first was the obvious suggestion that it was no more than a dictate of self-interest to support one's own community by fulfilling one's public duties: "Though all parties would rejoice their own principles prevailed, yet every one is more solicitous about his own safety than the others verity. Wherefore it cannot be unwise by the security of all, to make it the interest of all as well as the duty of all, to advance that of the public."[7] This thought was echoed by David Hume when he declared that "republican and free government would be an obvious absurdity, if [its characteristic institutions] made it not the interest, even of bad men, to act for the public good."[8]

But Penn also knew a subtler way of rationalizing individual selfishness as service to the common good. At first the argument was not fully thought out; it involved the notion that self-interest set free in the individual the greatest inventive and productive energies and the hope that selfish activities, if not beneficial to the whole, would be counterbalanced by similar actions of competitors. This discord of individuals, in other words, contributed to a concord of the whole. "And they are neither few," wrote William Penn, "nor of the weakest sort of men, that have thought the *concord of discords* a firm basis for government to be built upon. The business is to tune them well, and that must be the skill of the musician."[9] This phenomenon was frequently labeled *concordia discors* (or also, without change in meaning, *discordia concors*), a phrase borrowed from Latin antiquity and on one occasion translated as *Unanimity in Variance,* the title of an anonymous pamphlet of 1687.[10] *Concordia discors* designated quite generally the paradoxical harmony said to prevail in a community that tolerates disagreement, competition, and dissent among its individual members.[11]

Concordia discors was an engaging notion, but its functioning had not yet been explained. Some clarification of the concept was achieved, without referring to it by name, in Bernard de Mandeville's *Fable of the Bees* (1705/1714). The fable made a hyperbolic plea for the toleration not only of ordinary selfishness but even of behavior that was positively immoral. Inspired by his observations of life under the new constitutional government where the free interplay of political and economic forces was accompanied by a startling amount of corruption, usually unpublished, de Mandeville professed that "private vices" were "public benefits." Life in the beehives, he insisted, thrived as long as all vices were tolerated:

> The worst of all the multitude
> Did something for the common good.

This was the state's craft, that maintain'd
The whole, of which each part complain'd:
This, as in musick harmony,
Made jarrings in the main agree;
Parties directly opposite,
Assist each oth'r, as 'twere for spight;
And temp'rance with sobriety
Serve drunkenness and gluttony.
. . . .
Thus vice nursed ingenuity,
Which join'd with time and industry,
Had carry'd life's conveniencies,
It's real pleasures, comforts, ease,
To such height, the very poor
Lived better than the rich before.[12]

But as soon as all vice was stamped out, so the story ended, life in the bee-hive became not only dull but also very miserable.

Although the poem was received with cries of outrage, Mandeville had expressed a mood that was spreading. Economic historians have pointed to *The Fable of the Bees* as an early expression of the economic principle of laissez faire. Perhaps even greater was its importance as a statement of the liberal outlook in general. Its central theme was the turbulent harmony characteristic of life in a liberal society, a paradox summed up more concisely by Alexander Pope:

All Partial Evil, universal Good:
All Discord, Harmony, not understood.[13]

Pope's couplet expressed liberalism's answer to Hobbes's absolutist portrait of free society as a "war of all against all" in which "man was his own wolf." While the authoritarian attitude toward conflict was marked by fear, with persecution and prohibition expressing their response to opposition and dissent, in the liberal ideology, controversy, competition, and opposition were not only grudgingly tolerated but cheerfully welcomed as agents of a higher unity.

Pronouncements such as the preceding from Hooker to Pope express a new outlook on the functioning of human society. Those authors' advocacy of the free pursuit of self-interest, their insistence on the utility of toleration and compromise, and their faith in *concordia discors* were high-minded and forward-looking, but they appealed to the sentiment rather than to the intellect. To make the advocacy of their new values persuasive, a more solidly analytical argument was needed to show how liberty could produce order and how discord was to lead to harmony. Such an argument was supplied by the concept of self-regulation.

The concept of the self-regulating system had evolved, as was shown in the preceding chapters, in seventeenth- and eighteenth-century Britain in the discussion of several quite separate social problems. A self-regulating system has the characteristic ability to maintain its own stability alone—despite outside disturbances and without help from a higher authority—by virtue of its peculiar structure. The dynamic interactions of its elements form a circular cause-and-effect chain, constituting a closed or feedback loop that is saved from becoming a "vicious circle" by a sign change (which turns an increase into a decrease, and vice versa) of the signal during its travel once around the loop. In the supply-and-demand system, for example, the market acts upon the industry through the variable of the price and the industry upon the market, in turn, through that of the supply. Since the market responds to an *increase* in supply by a *decrease* in price (and vice versa), the effect of a random disturbance (resulting, say, in a price increase), after coming full circle, will be an appropriate countermeasure (a proportional decrease in price). Thus, the self-balancing capability of the system.

The system performs this feat without help from outside; all the system's elements are of equal rank, and equilibrium is achieved as an automatic outcome of their free interplay, not by the purposeful activity of a specialized agent. Therefore, the self-regulating system may be described as autonomous.

A few other characteristics of the self-regulating system are noteworthy. The system's interdisciplinary applicability has already become clear in previous discussions; more examples could be added indefinitely. Applications in eighteenth-century technology will be discussed later. The concept is universal because it can be reduced to an abstract pattern of cause-and-effect relationships; thus it lends itself splendidly to mathematical description.

Self-regulating systems, while autonomous, are capable of hierarchical superposition. One self-regulating system can function very well as a minor element within another higher or more general such system, and so forth. Thus it is possible to envision the entire universe as a network of superimposed and interacting self-regulating systems, maintaining themselves and the world permanently—despite occasional local lapses—in some sort of dynamic equilibrium.

In its fully developed state as an abstract, cybernetic concept, the notion of the self-regulating (or feedback) system is a product of the twentieth century. In eighteenth-century Britain the concept had not been recognized in its abstract universality nor, needless to say, did it have a name. The increasing frequency and clarity, however, with which the concept appeared in eighteenth-century British literature and with which it shaped important debates are a measure of the concept's popularity and of the power it held over the collective imagination. A significant portion of the British public embraced it as a general scheme for interpreting complex phenomena and for structuring complex systems. What won the concept such popularity was its promise of linking the values of equilibrium and liberty.

The idea of a system that would keep itself in balance without direction from higher authority was irresistible to the liberal mentality and would feed one of the basic liberal hopes: self-regulating systems are inherent almost everywhere in the natural and social world; in the face of problems, leave things alone—an optimal state will establish itself automatically! Belief in the beneficent functioning, the widespread existence, and the vast potential for further application of self-regulating systems was the basis for a liberal conception of order. Even if the functioning of such systems was clearly comprehended by perhaps only a minority, the values and attitudes deriving from an unspecific confidence in such systems were easily able to infect the entire public.

Practical life under the liberal conception of order exhibited a good deal less order than under its authoritarian alternative. There was less discipline, regularity, and predictability. Since the free pursuit of self-interest was believed to lead not only to the greatest individual happiness and prosperity but also to a maximum of creativity, energy, and initiative for the benefit of the whole society, this price was gladly paid. Confidence in the salutory functioning of the automatic system tended to go together with a distrust of authoritative action and of central planning (the ethos of "muddling through"); sometimes the overriding devotion to equilibrium had a strong resemblance with conservatism and dislike of change.

The liberal conception of order outlined here is a theoretical abstraction in the same manner as had been its authoritarian counterpart. While the authoritarian conception of order had a firm grasp on the collective mentalities of Continental European societies in the seventeenth and eighteenth centuries, the thinking of the British public fell more and more under the influence of this liberal alternative (which is not to say that practical life in Britain was consistently liberal in every aspect; consider only the spread of a thoroughly authoritarian form of factory management in the course of the Industrial Revolution).[14] These two abstractions seem useful in explaining certain social-intellectual phenomena in modern European history as long as they are used with an appropriate respect for the limitations and dangers of such constructs. They are not suggested to be mutually exclusive and collectively exhaustive. They are emphatically not intended as a single, simple dichotomous formula for explaining the history of the West. Their particular advantage seems to be that they offer additional elucidation without contradicting any accepted interpretations.

Although the liberal conception of order existed primarily in the human mind, its manifestations were observed in historical reality. As liberal principles and processes actually entered into the practical political and economic life in Britain, the sceptical observer was forced to admit that an order of sorts could be seen as establishing itself, an order vastly different from the discipline and regimentation customary under the authoritarian system, but a form of order nonetheless. Montesquieu, after a two-year stay in England, wrote about liberal society:

> What is called unity in the body politic is a very ambiguous matter: true unity is a harmonious one which accomplishes that all elements—however contrary they may appear to us—will contribute to the general good of society, as dissonances in music harmonize in the full chord. There can be unity in a state in which one seems to see nothing but discord. . . . It is the same as with the parts of the universe which are tied together eternally by the actions of the one and the reactions of the other.[15]

Edmund Burke, ostensibly commenting on an ancient French constitution, observed similarly:

> In your old states you possessed that variety of parts corresponding with the various descriptions of which your community was happily composed; you had all that combination, and all that opposition of interests, you had that action and counteraction, which in the natural and in the political world, from the reciprocal struggle of discordant powers, draws out the harmony of the universe.[16]

Karl Marx, finally, when commenting on the movements of prices subject to the market mechanism, put the same thing very briefly: "The overall movement of this disorder is its order."[17]

These three witnesses had recognized a form of order behind the unrestrained, turbulent spontaneity of life in a liberal society, and a tone of wonder at the reality of such a paradox is unmistakable in their testimony.

12

Self-regulating Mechanisms in Practical Technology

We still need to consider the relationship between the liberal conception of order and technology. In the various philosophical discussions reviewed above dealing with the emergence of the liberal corruption of order, there was no particular technological metaphor that stood out. One might conclude, on the face of it, that this conception of order was not influenced by technology; however, another kind of interdependence might have existed underneath the surface. What gave coherence to the liberal conception of order was the concept of self-regulation, and this is, above all, a technological concept. It seems appropriate to review the background of this concept in technology here.[1]

Classical antiquity knew self-regulating mechanisms in the form of liquid-level regulators like the float valves in automobile carburators and in bathroom water tanks.[2] Hero of Alexandria's *Pneumatics,* written in Greek in the latter half of the first century A.D. and first available to the European public in print and Latin translation in 1575, contains clear descriptions and good illustrations of several such devices (see figures 12–1 and 12–2).[3] Apparently the publication of this invention produced no reaction in the audience. While other ideas revealed in the rapidly distributed and widely read book were soon repeated in the literature and applied in practice, the level regulators, judging by the available sources, were ignored. They are first documented at the turn of the eighteenth century for use in regulating the levels of water supply tanks in city houses and of steam boilers (see figure 12–3).[4]

The first self-regulating system invented originally in modern Europe, it seems, was the thermostatic regulator of Cornelis Drebbel (1572–1633), a Dutch inventor and alchemist in the service of the English kings James I and Charles I. Drebbel's thermostat served to maintain constant temperatures in chicken incubators and in chemical furnaces where it was hoped to provide the key to transmutation (see

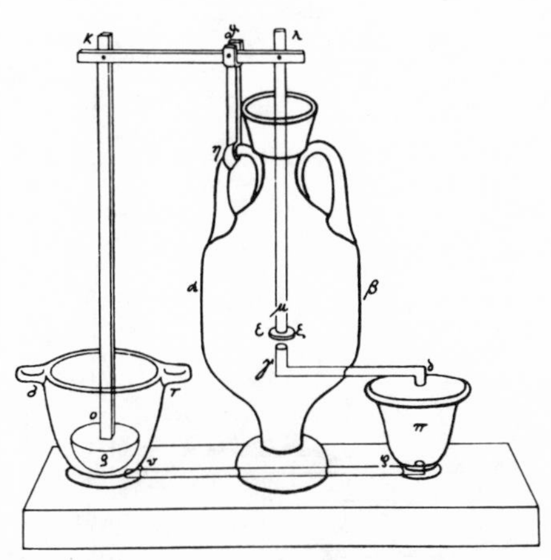

Figure 12–1. Float-level regulator by Hero of Alexandria, circa A.D. 60.

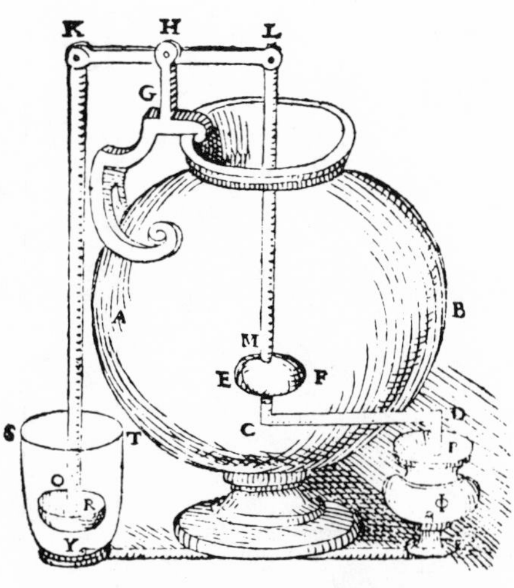

Figure 12–2. Hero's float-level regulator, as depicted in Federigo Commandino's 1575 Latin edition.

figures 12–4 and 12–5). The temperature in the space to be heated was measured by the expansion of alcohol vapor; this expansion was employed, via appropriate mechanical linkages, to adjust the air supply to the fire. Several contemporary descriptions of the device have survived which all indicate that it was actually built and was successful in practical operation. The scheme was employed again, presumably on the basis of those descriptions, by subsequent inventors and in time found its way into industrial use.[5]

New control problems arose with the invention of the steam engine. One of them was the regulation of the steam pressure in the boiler.[6] The simplest solution was blowing off excess steam through a safety valve, as was first proposed by Denis Papin in 1681 before the Royal Society of London (see figure 12–6). As an automatic control device, the safety valve, with its simple balance between the force of the steam pressure and an external weight, may appear trivial but can be shown to satisfy the formal criteria of the feedback principle. The safety valve was rapidly introduced into practice. A steam boiler without one would soon have been inconceivable. Inventions of several other, more elaborate pressure regulators in eighteenth-century Britain are also documented.

Much wider interest was aroused by the problem of regulating the speed of rotating machines, first of windmills and later of steam engines. An early solution was proposed in 1686 by G. W. Leibniz of Hanover,

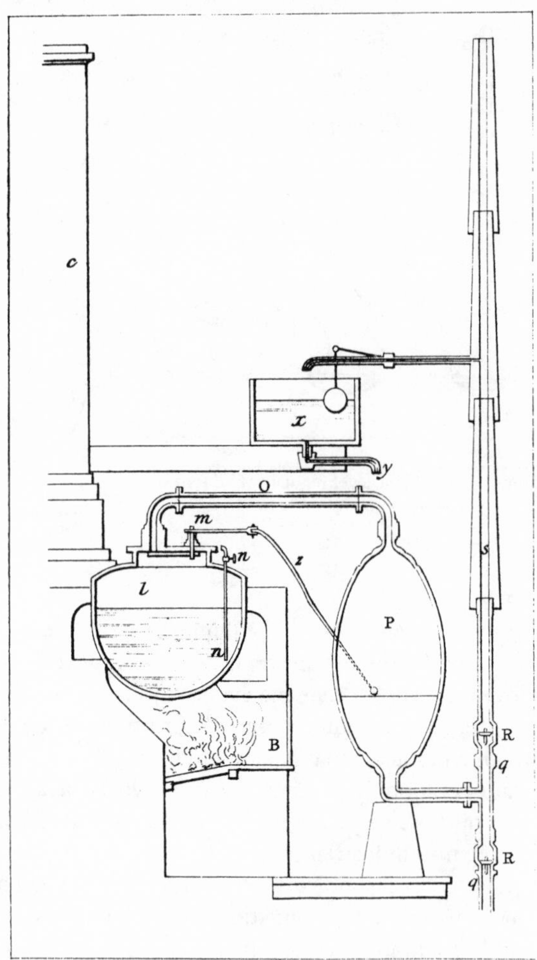

Figure 12–3. Float-level regulator in the feed-water supply tank of the steam pump by Thomas Savery, circa 1700.

Germany, a mechanical inventor better known as a philosopher, for rotating machines powered by wind, water, or weights (see figure 12–7).[7] In the principles of its action, the device was correctly conceived, but it was problematic in its mechanical details. The invention was sketched by hand on the back of a used envelope; there are no indications that it was ever tested in practice or communicated to others.

In the eighteenth century, several methods for regulating windmills were awarded English patents.[8] A simple way of avoiding excessive speed was to mount the windmill sails flexibly so that they could bend backward under a strong wind. In more fully articulated speed-control

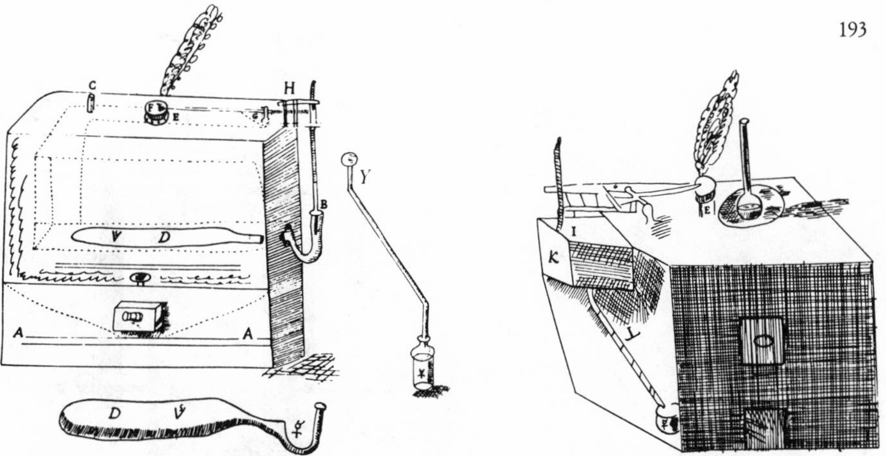

Figure 12–4. Cornelis Drebbel's thermostatically regulated chicken incubator, circa 1620. The glass tube, D, filled with alcohol in the large horizontal part (left) and with mercury in the U-tube (right), converts temperature increases in the furnace into upward motions of the rod, B, thus reducing the exhaust vent, E-F.

Figure 12–5. Cornelis Drebbel's thermostatically controlled alchemical furnace, circa 1620.

systems, a distinct speed-measuring device, for example, a centrifugal pendulum or a centrifugal fan blowing against a weighted baffle, would adjust the area or angle of the windmill sails in accordance with the speed (see figure 12–8).

Another self-regulating system associated with windmills that became popular in eighteenth-century Britain had the task of keeping the entire mill automatically pointed into the wind (see figure 12–9). This was accomplished by a small auxiliary wind wheel, mounted on the mill at a right angle to the main wheel and geared to a mechanism for turning the mill about its vertical axis. By virtue of its orientation, this wheel would be at rest as long as the mill was facing squarely into the wind. As soon as the mill deviated from the correct angle, the wheel would begin to rotate, with the effect of turning the mill into the wind. It would come to rest as soon as it was again parallel to the wind's direction.

Self-regulating devices such as these were described in British patent specifications and, in the course of the eighteenth century, became part of the standard practice of British millwrights. The invention of the best-known self-regulating mechanism of the eighteenth century, the steam engine governor, was a direct outgrowth of this windmill tech-

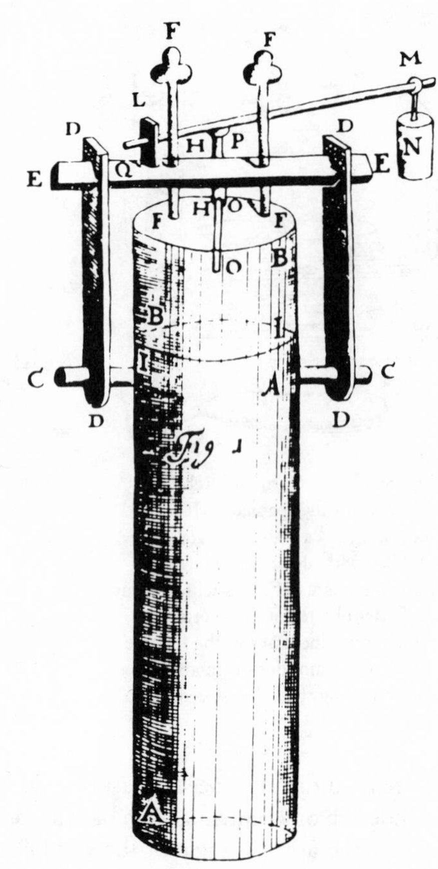

Figure 12–6. Pressure cooker with pressure regulator by Denis Papin, 1681.

nology. In a letter dated 28 May 1788, Matthew Boulton described to his associate James Watt a mechanism he had seen somewhere on a mill which incorporated a centrifugal pendulum as a speed-sensing device. Before the year was over, the first centrifugal governor was successfully installed on a Boulton-Watt steam engine, and within only a few years it had become an indispensable part of every stationary steam engine (see figure 12–10).[9]

The steam engine governor probably did more than any other agent to publicize the concept of self-regulation among engineers and the general population. The Watt steam engine was greeted as a machine of revolutionary importance and as the herald of a new age. No one would miss an opportunity to see this wonder in operation, and few who saw it would have failed to inquire about the purpose of those rapidly rotating centrifugal weights that were mounted conspicuously over the machine. To explain the concept of self-regulation, from that time on, one only

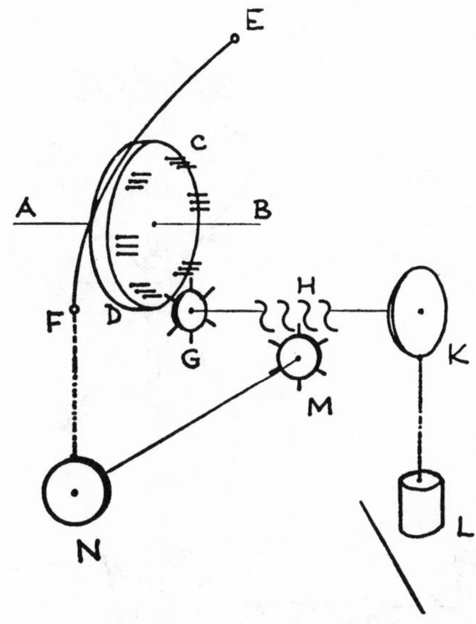

Figure 12–7. Speed regulation scheme for mills and so forth, by G. W. Leibniz, 1686. The brake, EF, acting on the fly wheel, CD, on the main shaft, AB, is tightened through the gear, G, and the worm gear, HM, at increasing speed and released by the counterweight, L, at decreasing speed.

had to point to the steam engine governor. When Norbert Wiener in 1947 christened his new science of cybernetics, he was expressly paying tribute to what he considered the earliest cybernetic device; the word *governor* is derived, via the Latin *gubernator,* from the Greek χυβερνήτης, steersman.[10]

The evidence, then, suggests that self-regulating devices were invented and introduced into practice in Britain earlier than anywhere else. To be sure, three early original inventions go to the credit of Continental Europeans (Drebbel, Papin, and Leibniz, although Drebbel and Papin were in England at the time), and other Continental inventions and applications followed. Nevertheless, a certain British preponderance in this technology seems undeniable.

The findings of this brief survey of the introduction of feedback mechanisms into modern technology are in curious agreement with the other results of this study: the acceptance of self-regulation into technology was closely contemporaneous with its introduction into social, political, and economic thought. And the role of Britain was as prominent in the one process as in the other.

Assuming that this observation is based securely on empirical evidence, and assuming that the similarity of patterns in the two developments was not a coincidence, one wonders how the two were connected. Thus, one must search for empirical evidence of links between the technological and the literary-philosophical uses of the concept of self-regulation. Did literary uses of the concept inspire technologists to

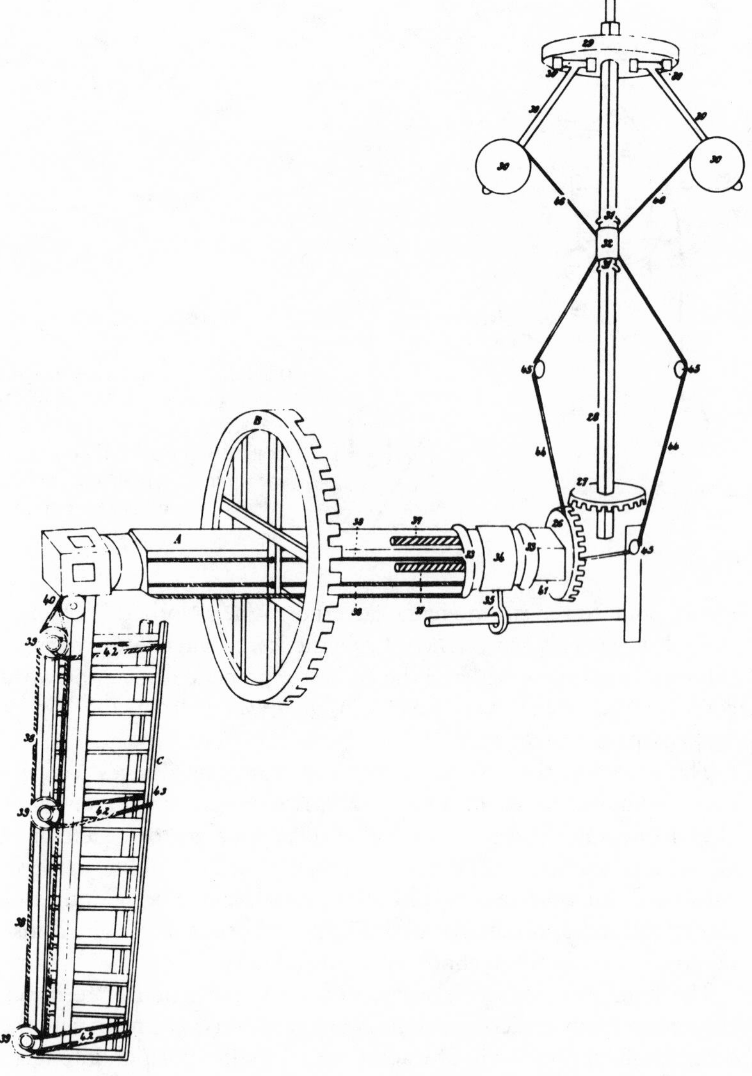

Figure 12–8. Speed regulator for windmills by Thomas Mead, 1787.

apply it to machinery? Or did writers introduce the notion of self-regulation into their arguments as a result of seeing it demonstrated on working machines? No contemporary testimony is known acknowledging any such direct influence in either direction.

It is tempting to recognize a similarity between the roles of the two mechanical analogies, that is, between the fondness of the authoritarian

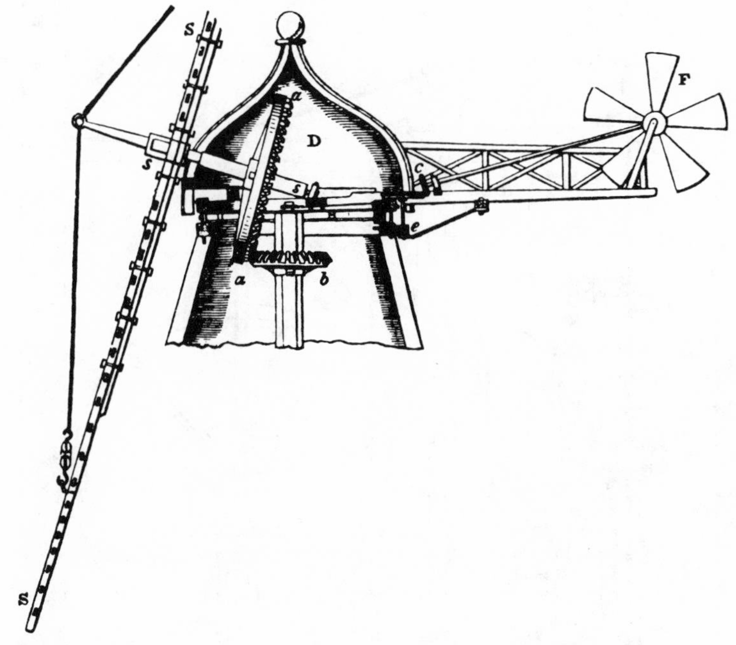

Figure 12–9. The "fan-tail."

conception of order for the mechanical clock and the affinity of the liberal conception of order for the notion of self-regulation. This similarity, however, has limits. In the sixteenth and seventeenth centuries, clockmaking had been a major industry. For centuries the mechanical clock had been discussed voluminously on all levels of literature; perhaps never in history has society been so fascinated with a single machine. The authoritarian conception of order was directly and patently shaped by society's experience with the mechanical clock.

The self-regulating mechanisms of the eighteenth century—that is, before the steam engine governor—by contrast, were few in number, inconspicuous in appearance, and unpublicized. By mid-eighteenth century, a person with technological interests who read moderately widely and was an alert observer might have seen self-regulating mechanisms in several forms, but probably without recognizing their common principle. By the same time, self-regulation as an abstract concept had already evolved in political and social discussions to a level of considerable maturity. It is hard to imagine that authors who offered particularly clear formulations of the concept—Gervaise, Hume, Smith, or, to a lesser extent, Halifax, Bolingbroke, or Blackstone—had been aided by a knowledge of actual feedback mechanisms. Moreover, if an author on,

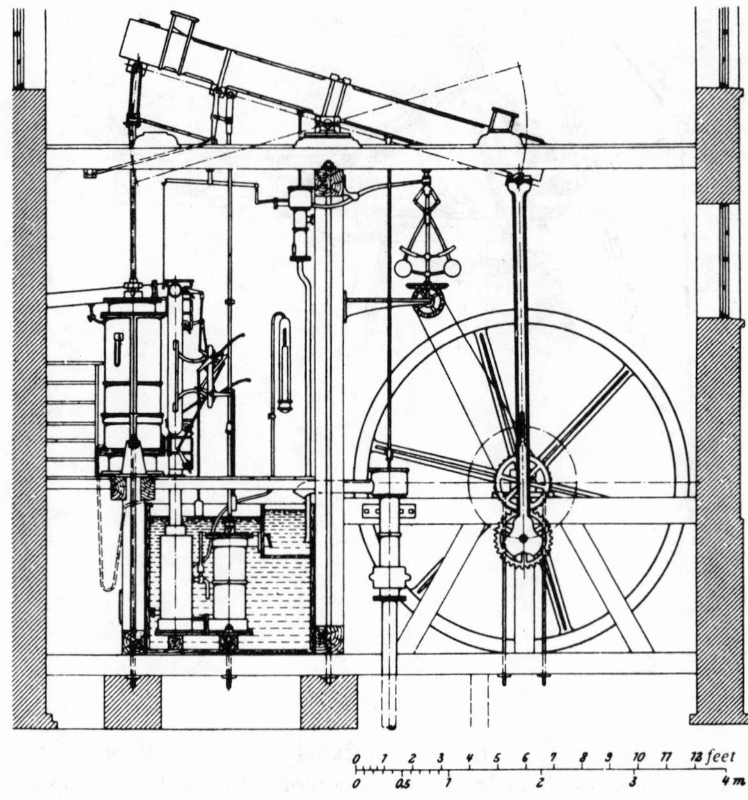

Figure 12–10. Centrifugal governor on a Watt steam engine, circa 1790.

say, economic theory in the mid-eighteenth century had recognized the analogy between a thermostatic device and the supply-and-demand mechanisms, would he not have been eager to record this insight? But no such testimony is known. One must conclude that the liberal conception of order was not significantly affected by the practical technology of self-regulating mechanisms.

As to influences in the opposite direction, one must ask whether technology was in some way affected by the spreading of the liberal concept of order. On the surface, there is no direct evidence of such influence. No contemporary testimony is known acknowledging that an inventor of a self-regulating device was inspired by philosophical uses of the concept. It must be admitted, however, that the nature of the surviving sources on such inventions makes it unlikely that a record of such an inspiration would have survived. Furthermore, what little we know about the nonverbal character of typical inventors makes it unlikely that they would have drawn direct inspiration from such sources in the first place.

At the same time, it seems extremely likely that the inventors were just as subject to the suggestive power of the liberal conception of order as their contemporaries and that this influence affected the character of their inventions and innovations. It may have made them appreciate the value of balance and equilibrium, and it may have taught them to consider the dynamic interactions between different parts of a system. They may have been encouraged to turn their ingenuity in directions that, without the new liberal attitudes, would not have been the case.

Another factor that shaped the difference in the fortunes of self-regulating mechanisms between Britain and the Continent was the response of the population. In Britain, self-regulating mechanisms were received more hospitably than on the Continent. The reasons for the warmer welcome were doubtless manifold, not excluding economic factors such as the greater demands and opportunities in the environment of the Industrial Revolution. But prominent among them was the following: if a technological innovation displays in structure and functioning an unmistakable analogy to the structure that a society prefers to give its various practical and theoretical systems, if it reflects the various mentalities and attitudes that shape public life, in short, if it matches and reinforces the prevailing conception of order, it will be received more warmly, regardless of its technical merits, than other inventions.

About the details of the causal nexus between the advent of the liberal conception of order and the rise of self-regulating mechanisms in technology we are reduced, at this point, to speculation. Quite firmly established, however, is the fact of the simultaneous appearance in Britain of these two phenomena, which in itself is forceful evidence of the interdependence of the socio-intellectual with the technological activities of a culture. The same interdependence is manifest again, for good measure, in the converse phenomena on the Continent, namely, in the symbiotic relationship of the authoritarian conception of order with the mechanical clock and its attendant disfavor with self-regulation. Technology, thus, is part of the intellectual life of a culture.

Notes

Translations, when not otherwise noted, are by the author.

Introduction

1. Otto Mayr, *The Origins of Feedback Control* (Cambridge, Mass., 1970), pp. 125–31.

2. Ibid., pp. 46–48.

3. Otto Mayr, "Adam Smith and the Concept of the Feedback System," *Technology and Culture* 12 (1971): 1–22.

Chapter 1. The Mechanical Clock, Its Makers and Users

1. For comprehensive discussions of medieval technology see, for example, Friedrich Klemm, *Technik: Eine Geschichte ihrer Probleme* (Munich, 1954), pt. 2; Lynn White, Jr., *Medieval Technology and Social Change* (Oxford, 1962); Jean Gimpel, *The Medieval Machine* (New York, 1976); and Lynn White, Jr., *Medieval Religion and Technology* (Berkeley and Los Angeles, 1978).

2. Much has been written on the origins of the mechanical clock. Representative recent works are C. F. C. Beeson, *English Church Clocks, 1280–1850* (London and Chichester, 1971) and John D. North, ed. and trans., *Richard of Wallingford,* 3 vols. (Oxford, 1976). Although the evidence presented there points to England, one wonders whether further research might not bring to light evidence of similarly early clocks in France, Germany, Italy, and elsewhere. For older discussions, see also Ernst Zinner, *Aus der Frühzeit der Räderuhr, Deutsches Museum, Abhandlungen und Berichte,* vol. 22.3 (Munich, 1954); Guiseppe Boffito, "Dove e quando potè Dante vedere gli orologi meccanici che descrive in *Par.* x, 139; xxiv, 13; xxxiii, 144?" *Giornale Dantesco* 39 (1936): 45–61; and A. P. Usher, *A History of Mechanical Inventions* (Cambridge, Mass., 1954), chap. 8.

3. On the social background of medieval timekeeping, see, for example, Gustav Bilfinger, *Die mittelalterlichen Horen und die modernen Stunden* (Stuttgart, 1892); Richard Glasser, *Time in French Life and Thought* (1936), trans., C. G. Pearson (Manchester and Totowa, N.J., 1972); Jacques Le Goff, "Au Moyen âge: Temps de l'église et temps du marchand," *Annales E.S.C.* 15

(1960): 417–33; Carlo M. Cipolla, *Clocks and Culture, 1300–1700* (New York and London, 1967); John D. North, "Monasticism and the First Mechanical Clocks," in J. T. Fraser and N. Lawrence, eds., *The Study of Time II,* Proceedings of the Second Conference of the International Society for the Study of Time (New York, 1975), pp. 381–98; and David S. Landes, *Revolution in Time: Clocks and the Making of the Modern World* (Cambridge, Mass. and London, 1983), esp. chaps. 3 and 4.

4. George Sarton, *A History of Science,* 2 vols. (Cambridge, Mass., 1952), 1:30.

5. Ibid., p. 75.

6. For a brief summary of the early history of the water clock, see Usher, *History of Mechanical Inventions,* pp. 187–192.

7. Derek J. de S. Price, *Gears from the Greeks* (New York, 1975).

8. Derek J. de S. Price, "On the Origin of Clockwork, Perpetual Motion Devices, and the Compass," *U.S. National Museum Bulletin 218* (Washington, D.C., 1959), pp. 81–112; John D. North, "Opus quorundam rotarum mirabilium," *Physis* 8 (1966): 337–72.

9. Examples of nontraditional timekeepers from the thirteenth century are found in the notebook of Villard de Honnecourt (ca. 1235), H. R. Hahnloser, ed., *Villard de Honnecourt* (Vienna, 1935), pl. 12; in the *Libros del saber de astronomia* of King Alfonso X of Castile, ed. M. Rico y Sinobas (Madrid, 1866), 4: 67–76; and in C. B. Drover, "A Medieval Monastic Water-Clock," *Antiquarian Horology* 1.5 (1954): 54–58, 63, and 84.

10. L. Thorndyke, "Invention of the Mechanical Clock about 1271 A.D.," *Speculum* 16 (1941): 242–43; see also White, *Medieval Technology,* p. 122.

11. For the former argument, see Joseph Needham, Wang Ling, and Derek J. Price, *The Heavenly Clockwork* (Cambridge, 1960). For the latter argument, see the authors cited in note 2. The case against the Needham-Ling-Price thesis is well summarized in Landes, *Revolution in Time,* chap. 1.

12. North, *Richard of Wallingford;* Alfred Ungerer, *L'Horloge astronomique de la cathédrale de Strasbourg* (Paris, 1922); Silvio A. Bedini and Francis R. Maddison, "Mechanical Universe: The Astrarium of Giovanni de' Dondi," *Transactions of the American Philosophical Society,* n.s. 56.5 (1966).

13. For standard summaries of this development, see Usher, *A History of Mechanical Inventions,* chap. 12, and, more recently, Klaus Maurice, *Die deutsche Räderuhr,* 2 vols. (Munich, 1976), 1:81–126. On its origins, see J. H. Leopold, ed., *The Almanus Manuscript* (London, 1971).

14. A comprehensive treatment of German clockmaking in the period 1550–1650 is Klaus Maurice and Otto Mayr, eds., *The Clockwork Universe: German Clocks and Automata, 1550–1650* (Washington, D.C. and New York, 1980); see also Maurice, *Die deutsche Räderuhr,* passim.

15. Conrad Dasypodius, *Warhafftige Ausslegung des Astronomischen Uhrwercks zu Strassburg* (Strasbourg, 1578), preface, p. 7.

16. Samuel Macey, *Clocks and the Cosmos: Time in Western Life and Thought* (Hamden, Conn., 1980).

17. Adam Smith, *The Wealth of Nations,* ed. E. Cannan, Modern Library Edition (New York, 1937), p. 243.

18. An example is Christianopolis, the utopian community of the Rosicrucians, as described by Johann Valentin Andreae in *Die chymische Hochzeit Christiani Rosenkreutz Anno 1459* (1616) (Munich, 1957), p. 98.

19. Johannes Geyger, the former author, says:

Ein richtig Uhrwerck in der Stadt/
Zeigt an daß da ein weiser Raht/
Ein richtigs Regiment führ eben/
Auch gute Policey darneben/
Die Bürger regier mit Weißheit/
Ertheil nach Grechtigkeit die Bscheid.

Geyger, *Horologium politicum* (Nuremberg, 1621), p. 39.

According to the latter author, "If we lived without clocks we would be like unreasoning beasts that distinguish the hours only by waking or sleeping" ("Sollten wir ohne Uhren leben, so würden wir gleich sein dem unvernünftigen Viehe, welches die Stunden mit dem schlaffen und wachen unterscheidet") (Georg Philipp Harsdörffer, *Delitiae mathematicae et physicae,* pt. 2 [Nuremberg, 1651], pt. 319).

20. Iorwerth C. Peate, *Clock and Watch Makers in Wales* (Cardiff, 1975), p. 14.

21. François Rabelais, *Five Books of the Lives, Heroic Deeds, and Sayings of Gargantua and his Son Pantagruel,* trans. Thomas Urquhart and Peter Antony Motteux, 3 vols. (London, 1964), 1:114, 139.

22. Théodore Agrippa d'Aubigny, *Adventures du Baron Faeneste* (1617), quoted from Henri Michel, "L'Horloge de sapience et l'histoire de l'horlogerie," *Physis* 2 (1960): 291–98.

23. To offer a few random samples, Johann Valentin Andreae called his chapter on the mechanical clock "Authomatica," as distinct from the preceding chapter "Gnomonica," on sundials (Johann Valentin Andreae, *Collectanea mathematica decades XI* [Tübingen, 1614]). Geyger simply identified automata and clocks: "Automatum, das ist ein solches künstliches, selbstgehendes und schlagendes Uhrwerck" (Geyger, *Horologium politicum,* p. 25). Similarly, see Christian Wolff: "Automaton heisset eine Maschine, die entweder durch Gewichte, oder durch eine Feder beweget wird, solchergestalt daß es das Aussehen hat, als wenn sie sich selbst bewegete. Nemlich die bewegende Kraft ist mit ein Theil derselben. Dergleichen sind die Uhren und Bratenwender" (*Mathematisches Lexicon* [Leipzig, 1716], col. 211), and John Harris: "Automata are Mechanical or Mathematical Instruments or Engines, that going by a Spring, Weight, etc. seem to move of them selves, as a Watch, Clock, etc." (*Lexicon Technicum,* 2 vols. [London, 1736], vol. 1, unpaginated).

24. Lynn White, Jr., "Medical Astrologers and Late Medieval Technology," in White, *Medieval Religion and Technology,* pp. 297–316.

25. On astronomical clocks, see Henry C. King, *Geared to the Stars: The Evolution of Planetariums, Orreries, and Astronomical Clocks* (Toronto, 1978); Anthony J. Turner, *The Clockwork of the Heavens,* catalog of an exhibition presented by Asprey & Co. (London, 1973); and Maurice and Mayr, *The Clockwork Universe,* chap. 5.

26. See Maurice and Mayr, *The Clockwork Universe,* chap. 4. On the history

of automata in general, see also Alfred Chapuis and Edmond Droz, *Automata: A Historical and Technological Study,* trans. Alec Reid (Neuchâtel and New York, 1958); Derek J. de S. Price, "Automata and the Origins of Mechanism and Mechanistic Philosophy," *Technology and Culture* 5 (1964): 9–23; and Silvio A. Bedini, "The Role of Automata in the History of Technology," *Technology and Culture* 5 (1964): 24–42.

27. Alex Keller, "Mathematical Technologies and the Growth of the Idea of Technical Progress in the Sixteenth Century," in Allen G. Debus, ed., *Science, Medicine, and Society in the Renaissance: Essays to Honor Walter Pagel,* 2 vols. (New York, 1972), 1:17.

28. D. Bruce, "Human Automata in Classical Tradition and Medieval Romance," *Modern Philology* 10 (1912–13): 511–26; M. Sherwood, "Magic and Mechanics in Medieval Fiction," *Studies in Philology* 44 (1947): 567–92; Otto Mayr, "Automatenlegenden in der Spätrenaissance," *Technikgeschichte* 41 (1974): 20–32; and William Eamon, "Technology as Magic in the Late Middle Ages and the Renaissance," *Janus* 70 (1983): 171–212.

Chapter 2. The Rise of the Clock Metaphor

1. For the sake of simplicity, the word *metaphor* will be used here as a general name for all the various types of literary comparisons and figurative expressions such as simile, parable, analogy, allegory, and so forth.

2. *Automata* are included here under the category *clock,* although at the time under review automata were considered the higher category.

3. Notably, Lynn White, Jr., in his *Medieval Technology and Social Change* (Oxford, 1962), p. 124, n. 3. Recent evidence (see chap. 1, n. 2) shows that the mechanical clock was invented about half a century earlier than White had assumed.

4. Guiseppe Boffito, "Dove e quando potè Dante vedere gli orologi meccanici che descrive in *Par.* x, 139; xxiv, 13; xxxiii, 144?" *Giornale Dantesco* 39 (1936): 45–61.

5. Dante Alighieri, *Divine Comedy,* trans. H. R. Huse (New York and Toronto, 1954), *Paradiso* x, 139–48.

6. Ibid., *Paradiso* xxiv, 10–18.

7. Lynn White, Jr., "The Iconography of Temperantia and the Virtuousness of Technology," in T. K. Rabb and J. E. Seigel, eds., *Action and Conviction in Early Modern Europe* (Princeton, N.J., 1969), p. 211. For the following, I am greatly indebted to this excellent article.

8. According to the *Lexikon for Theologie and Kirche,* 2d ed., 14 vols. (Freiburg, 1958), vol. 2, Berthold of Freiburg is not identical with "Berthold Huenlein."

9. "Volo autem hunc libellum horologium devotionis intitulare hac ratione: Nam sicut dies naturalis habet xxiv horas, diem et noctem simul computando, sic iste libellus de vita Christi habet xxiv capitula . . . et quodlibet capitulum huius libri duxi horam nominandam" (Dominikus Planzer, "Gab es eine gekürzte Redaktion des lateinischen Horologium sapientiae des sel. Heinrich Seuse O. P.?" *Divus Thomas* 13 (1935): 208).

10. On Suso, see Dominikus Planzer, "Zur Textgeschichte und Textkritik des Horologium sapientiae des sel. Heinrich Seuse O. P.," *Divus Thomas* 12 (1934): 129–64, 257–78; Jeanne Ancelet-Hustache, "Quelques indications sur les manuscrits de l'Horloge de Sapience," in E. M. Filthaut, ed., *Heinrich Seuse: Studien zum 600. Todestag, 1366–1966* (Köln, 1966), pp. 161–70; and also White, "The Iconography of Temperantia" (see n. 7, above), pp. 207, 210, and 211–12.

11. White, "The Iconography of Temperantia," p. 211. The original Latin text is: "Unde et presens opusculum in visione quadam sub cuiusdam horologii pulcherrimi, rosis [*sic*—error for rotis] speciosissimis decorati, et cymbalorum bene sonancium et suavem ac celestem sonum reddencium, cunctorumque corda sursum movencium varietate perornati figura, dignata est ostendere clemencia salvatoris" (Henricus Suso, *Horologium sapientiae,* ed. Joseph Strange [Cologne, 1861], pp. 9–10).

12. See H. Michel, "Some New Documents in the History of Horology," *Antiquarian Horology* 3 (1962): 228–91; H. Michel, "*L'Horloge de sapience* et l'histoire de l'horlogerie," *Physis* 2 (1960): 291–98; and Eleanor P. Spenser, "*L'Horloge de sapience:* Bruxelles, Bibliothèque Royale, MS IV, 111." *Scriptorium* 17 (1963): 277–99. The contents of these articles are summarized in White, "The Iconography of Temperantia," passim.

13. *Gesamtkatalog der Wiegendrucke* (Leipzig, Stuttgart, Berlin, 1925), vol. 4, col. 51–57.

14. Planzer, "Zur Textgeschichte und Textkritik" (n. 10, above), p. 130.

15. Jean Froissart, *Oeuvres: Poésies,* ed. A. Scheler, 3 vols. (Paris, 1870–72). For critical discussions of Froissart's poetic efforts, see also *Oeuvres de Froissart,* ed. Kervyn de Lettenhove, 25 vols. (Brussels, 1867–77), 1:99, 111; and F. S. Shears, *Froissart, Chronicler and Poet* (London, 1930), esp. pp. 202–4.

16. Froissart, ed. Scheler, 1:53–86.

17. The lead with Beauty doth full well accord;
Plesaunce by the cord is typified
So truly that it might not better be expressed;
For just as the weight draws
The cord to itself, and the taut cord
(When the cord is right well strained)
Pulls it towards itself and makes it move,
Which otherwise would never move,
Thus Beauty draws towards itself, and wakes
The Plesaunce of the heart.

Lines 113–22, from a partial translation in J. Drummond Robertson, *The Evolution of Clockwork* (London, 1931), p. 55.

18. Robertson, p. 57, lines 201–20.

19. This issue is discussed in detail in White, "The Iconography of Temperantia," pp. 203–4. The following discussion is largely based on White's article.

20. From the British Museum, Harley MS 4331, and Bibliothèque Nationale, MS fr. 606, quoted from White, "The Iconography of Temperantia," p. 209, who in turn has quoted from Rosemond Tuve, "Notes on the Virtues and Vices," *Journal of the Warburg and Courtauld Institutes* 24 (1963): 289.

21. See n. 12, above.

22. White, "Iconography of Temperantia," pp. 213–17.

23. "Qui a lorloge soy regarde / En tous ses faicts heure et temps garde. / Qui porte le frain en sa bouche / Chose ne dist qui a mal touche. / Qui lunettes met a ses yeux / Pres lui regarde sen voit mieux. / Esperons monstrent que cremeur [i.e., fear] / Font estre le josne homme meur. / Au moulin qui le corps soutient / Nuls exces faire nappartient" (Bibliothèque Nationale, MS fr. 9186, in E. Mâle, *L'Art religieux de le fin du moyen âge en France,* 4th ed. [Paris, 1931], pp. 311–16). Quoted from White, "Iconography of Temperantia," p. 214.

24. Nicole Oresme, *Tractatus de commensurabilitate vel incommensurabilitate motuum celi,* ed. Edward Grant, (Madison, Wisc., 1971). See also Edward Grant, "Nicole Oresme and the Commensurability or Incommensurability of the Celestial Motions," *Archive for the History of Exact Sciences* 1 (1961): 420–58.

25. "Nam et si quis faceret horologium materiale nonne efficeret omnes motus rotasque commensurabiles iuxta posse? Quanto magis hoc opinandum est de architectore illo qui omnia fecisse dicitur numero, pondere, et mensura? Nulla autem incommensurabilia sunt numeris mensurata" (Oresme, *Tractatus de commensurabilitate,* pp. 292–95).

26. Nicole Oresme, *Le Livre du ciel et du monde,* ed. A. D. Menut and A. J. Deverny, trans. A. D. Menut (Madison, Wisc., 1968), p. 289.

27. Titus Lucretius Carus, *De rerum natura libri sex,* 5.96, ed. Cyril Bailey, 3 vols. (Oxford, 1947), 1:436.

28. Lactantius, *Divinae institutiones,* 2.5.13 (ca. A.D. 300), cited in Klaus Maurice, *Die deutsche Räderuhr* (Munich, 1976), 1:46; Arnobius Afer, *Adversos nationes,* 1.2 (ca. A.D. 300), cited in White, *Medieval Technology,* p. 174, n. 5; Firmicus Maternus, *De errore profanorum religionum* (ca. A.D. 350), ed. K. Ziegler (Munich, 1953), p. 76, line 26; Dionysius Areopagita, cited without source by White, *Medieval Technology,* p. 174, n. 5; Chalcidius, *Timaeus a Calcidio translatus commentarioque instructus* (ca. A.D. 400), in *Plato latinus,* ed. J. H. Waszink (London and Leiden, 1962), vol. 4, p. 184, chap. 147 and p. 301, chap. 299; Cassiodorus Senator, *Variae,* liber 2, epistula 40 and liber 5, epistula 42 (ca. A.D. 520), in *Opera* (Paris, 1588), 1:35a, 92a; Alanus ab Insulis, *The Complaint of Nature* (De planctu naturae, ca. 1150), trans. D. M. Moffat (New York, 1908), p. 25; John Sacrobosco in Lynn Thorndike, *The Sphere of Sacrobosco and Its Commentators* (Chicago, 1949), pp. 78, 117, 119, 142; and Robert Grosseteste, *De Sphaera,* chap. 1, lines 1, 5, and 12–13, in *Die philosophischen Werke,* ed. Ludwig Baur, (Münster, 1912), 9:11.

29. Marcus Tullius Cicero, *De natura deorum,* liber 2, 35, trans. H. Rackham (Cambridge, 1933), pp. 207–17, cited in Klaus Maurice, *Die deutsche Räderuhr,* 2 vols. (Munich, 1976), 1:46.

30. Clemens Romanus, *Recognitiones* 8.20.6–8, cited in Maurice, *Die deutsche Räderuhr,* 1:46.

31. The following is based on Nicholas H. Steneck, *Science and Creation in the Middle Ages: Henry of Langenstein (d. 1397) on Genesis* (Notre Dame, Ind., 1976).

32. Ibid., pp. 92, 112, 118, 149.

33. This might be the place to mention a few other early clock metaphors that have not been discussed previously:

—An extremely early one appears in Thomas Aquinas (1225?–1274), *Summa theologiae,* ed. Thomas Gilby (New York, 1970), 17:128–29; I have not discussed it because it is too vague in content and too early for the mechanical clock.
—A comparison of the heavenly spheres, as propelled by angels, with clockwork is offered in Giovanni da Fontana (1395?–1455?), *Liber de omnibus rebus naturalibus quae continentur in mundo* (ca. 1450) (Venice, 1544), fol. 41 r (2.9); cited in Lynn Thorndike, *A History of Magic and Experimental Science,* 5 vols. (New York, 1934), 4:169.
—Nikolaus von Cues (ca. 1400–64), *Von Gottes Sehen. De Visione Dei,* trans. E. Bohnenstaedt (Leipzig, 1942), pp. 86–87, introduces several clock metaphors that I found too abstract to be meaningful.
—Gaspare Visconti (1461–99) wrote several love poems based on clock analogies which Carlo M. Cipolla (*Clocks and Culture 1300–1700* [New York and London, 1967], p. 105) has compared favorably with those by Froissart. Their poetic quality may be higher, but they are less specific in their mechanical imagery (Rodolfo Renier, "Gaspare Visconti," *Archivio Storico Lombardo,* 2d ser. 13 [1886]: 540–45).

34. For a comprehensive treatment, see Arthur Henkel and Albrecht Schöne, eds., *Emblemata: Handbuch zur Sinnbildkunst des 16. and 17. Jahrhunderts* (Stuttgart, 1967).

35. Filippo Picinelli, *Mundus symbolicus,* Latin trans. by Augustinus Erath, 2 vols. (Cologne, 1687), 2:182–95; also see index s.v. *horologium.*

36. Erwin Panofsky, "Father Time," in Panofsky, *Studies in Iconology: Humanistic Themes in the Art of the Renaissance* (New York, 1939), pp. 69–93.

37. Surveys of such material can be found in Erwin Stürzl, *Der Zeitbegriff in der Elisabethanischen Literatur: The Lackey of Eternity* (Vienna and Stuttgart, 1965); T. H. Svartengren, *Intensifying Similes in English* (Lund, 1918); and W. Vosskamp, *Untersuchungen zur Zeit- und Geschichtsauffassung im 17. Jahrhundert bei Gryphius und Lohenstein* (Bonn, 1967).

38. Tusser, *Husb.* (1573), p. 4, quoted from Svartengren, *Intensifying Similes,* p. 371; Thomas Nashe, *Christs Teares over Jerusalem* (1593), 11:94, quoted from Svartengren, *Intensifying Similes,* p. 372; "eine wolgefaste Ordnung, Uhr und Richtschnur," Johann Conrad Dannhauer, *Catechismusmilch,* 2 vols. (1657), quoted from Jacob and Wilhelm Grimm, *Deutsches Wörterbuch* (Leipzig, 1854–1960), 11.2:737; and François Rabelais, *Five Books of the Lives, Heroic Deeds, and Sayings of Gargantua and his Son Pantagruel,* trans. Thomas Urquhart and Peter Antony Motteux, 3 vols (London, 1964), 1:128.

39. We see it euident in automaticall instruments, as clockes, watches, and larums, how one right and straight motion, through the aptnes of the first wheele, not only causeth circular motion in the same, but in diuers others also: and not only so, but distinct in pace, and time of motion: some wheeles

passing swifter than other some, by diverse rases: nowe to these deuises, some other instrument added, as hammer and bell, not only another right motion springeth thereof, as the stroke of the hammer but sound also oft repeated, and deliuered at certaine times by equall pauses, and that either larum or houres according as the partes of the clocke are framed. To these if yet moreover a directory hand be added, this first, & simple & right motion by weight or straine, shall seme not only to be author of deliberate sound and to counterfet voice, but also to point with the finger as much as it hath declared by sound. Besides these we see yet a third motion with reciprocation in the ballance of the clock. So many actions diuerse in kinde rise from one simple first motion, by reason of variety of ioynts in one engine. If to these you adde what wit can deuise, you may find all the motion of heaven with his planets counterfetted, in a small modil, with distinction of time and season, as in the course of the heauenly bodies. (Timothy Bright, *A Treatise on Melancholy* [London, 1586], p. 45, from Anthony J. Turner, *The Clockwork of the Heavens,* catalog of an exhibition presented by Asprey & Co. [London, 1973], p. 16)

40. John Donne, Sermon 53, *The Works,* ed. Henry Alford, 6 vols. (London, 1839), 2:492, quoted from M. A. Rugoff, *Donne's Imagery: A Study in Creative Sources* (New York, 1939), p. 117.

41. Though the very habit be but a ceremony, yet the distinction of habits is rooted in nature, and in morality; and when the particular habit is enjoined by lawful authority, obedience is rooted in nature, and in morality, too. In a *watch,* the string moves nothing, but yet it conserves the regularity of the motion of all. Ritual, and ceremonial things move not God, but they exalt that devotion, and they conserve that order, which does move Him. (John Donne, Sermon 122, *Works,* 5:175, from Rugoff, *Donne's Imagery,* p. 117).

42. John Amos Comenius, *The Great Didactic* (probably completed in 1632), trans. and ed. M. W. Keatinge (London, 1896), p. 249.

43. Ist nach dem *supremo Gubernatori, Directori,* Regierer und Führer aller ding auff Erden / den himmlischen Liechtern / Zeichen und Planeten / auch seinem heiligen Wort / sampt andern nützlichen Ordinantzen, Statuten und Gesatzen der weltlichen Christlichen Oberkeit / der rechte *Principal, Moderator,* Regierer und Führer deß menschlichen Lebens, und desselben gantzen Handels und Wandels. (Johannes Geyger, *Horologium politicum* [Nuremberg, 1621], pp. 44–45)

44. Antonio de Guevara, *The Diall of Princes,* trans. Thomas North (London, 1557; reprinted, Amsterdam, 1968), prologue.

45. John Webster, *The White Devil* (1612), in *The Complete Works,* ed. F. C. Lucas, 4 vols. (New York, 1937), 1:120.

46. Ein Fürst und Regent ist daß Lands Uhr
jeder richt sich nach denselben in Wercken
als wie nach der Uhr in Geschäfften.

Christoph Lehmann, *Florilegium politicum: Politischer Blumengarten* (1630; Frankfurt, 1662), p. 693.

47. William Davenant, *Poem, Upon His Sacred Majesty's Most Happy Return* (London, 1660), from Ruth Nevo, *A Dial of Virtue: A Study of Poems on Affairs of State in the Seventeenth Century* (Princeton, N.J., 1963), p. 145.

48. Cesare Ripa, *Iconologia* (1618), from I. Baudoin, *Iconologie* (Paris, 1644), 1:162, cited in Maurice, *Die deutsche Räderuhr,* 1:15.

49. John Donne, *The Poems,* ed. Herbert J. C. Grierson, 2 vols. (1912; London, 1929), 1:275–76.

50. Dieses Uhrwerck [including indications of the day and night hours, course of the planets, phases of the moon, alarm, strikes of the hours and quarter hours, etc.] betrachtete der glückselige Gleichnis-Erfinder Lipsius und sagte:

> Wie wir den Zeiger auf der Uhr sehen / und die Stunden aus seinem Umblauff erkennen / den Kunstrichtigen Gang aber / der ineinandergehaspelten Rädlein nicht verstehen; also erkennen wir zwar Gottes Gnaden und Strafzeichen / deroselben geheime Ursachen aber / wissen und verstehen wir nicht; wie auch der Fürsten und Herrn thun uns für Augen liget / ihre Rathschläge aber / und was sie darzu beweget / ist für unsern Augen verborgen. (Georg Philipp Harsdörffer, *Delitiae mathematicae et physicae* [Nuremberg, 1651], pp. 348–49)

51. "Gottes Uhr gehet gewiß und fehlet nicht / darvor hilfft kein Weißheit noch Gewalt. Wer etwas thun will / der sehe auff Gottes Uhr / versuche es / ob die Stund geschlagen / darinn er sein Geschäfft verrichten will" (Lehmann, *Florilegium politicum,* p. 935).

52. "Welche der Geist Gottes treibet / und sie verstehe / ihm auch folgen / die sind Kinder Gottes / wie das Uhrwerck / so es anderst recht gemacht ist / dem Gewichte gerne folget" (Geyger, *Horologium politicum,* p. 72).

53. "Der Puls ist in dem Menschen die Unruhe / und wie die Uhr ohne solche nicht mehr gehen kan / also hat deß Menschen Leben mit dem Schlagen der Pulsader seine Endschafft" (Harsdörffer, *Delitiae mathematicae,* p. 347).

54. "Weilen auch für das ander das materialische Uhrwerck von Rädern / Unruhe / Gewicht / Federn / Zeigern / etc. ein sehr wunderbar Werck ist: Als wird auch bei solchem die *mirabilis structura* und *fabricatio,* das ist / das wunderbare Gebäu des menschlichen Leibs nicht unfüglich betrachtet" (Geyger, *Horologium politicum,* p. 66).

55. Luis de Granada, *An Excellent Treatise of Consideration and Prayer* (anonymous translation of *Libro de la oración y meditación,* Salamanca, 1573), (London, 1599), p. 81. I owe this quotation as well as the following to Francis C. Haber, University of Maryland.

56. Thomas Tymme, *A Silver Watch-Bell* (London, 1614), p. 24.

57. Donne, "A Funerall Elegie" (1610), in *The Poems,* 2:37–40.

58. As when a curious clock is out of frame
a workman takes in peeces small the same

and mending what amisse is to be found
the same reioynes and makes it trewe and sound
so god this ladie into two parts tooke
too soon her soule her mortall corse forsooke
But by his might att length her bodie found
shall rise reioyned unto her soule now cround
Till then they rest in earth and heaven sundred
att which conioyened all such as live then wondred.

Epitaph on the tomb of Lady Dodderidge, who died 1 March 1614, Exeter Cathedral, from Turner, *The Clockwork of the Heavens,* p. 26.

The funeral poem is entitled "Die Lebens-Uhr":

Man mag die Lebens-Uhr / wohin man will verdrehen / Doch giebt sie keine Ruh / ihr Stand ist Unbestand; Die Unruh lässet sich wol fühlen: wo nicht sehen: So richtet unsern Lauf des höchsten Künstlers Hand. Offt schweigt der Herzens-Schlag / der Himmels-Zeiger stecket / indem der faule Rost den freyen Paß verschleusst; bis eine Kreuzes-Last die Räder wieder wecket / und Gott sein Gnaden-Oel in alle Fugen geusst. Indessen bleibt es fest / daß Hoffnungs-Circuln wanken; Du hast uns den Compaß / O Floridan / verrückt! Durch deinen frühen Fall entstehen die Gedanken: Dein Tod und Leben wird aus einem Bild erblickt. Ob manche falsche Uhr nicht zeiget / wie sie schläget / Doch traffen Gang und Klang in deiner redlich ein: Du hast den Ausspruch wol gedreht und überleget / Drum konte auch hernach dein Thon gewichtig seyn. Dich hat der scharfe Zahn des Sorgen-Rads benaget / bis dich die schwere Last der Krankheit überwug; und jene Stunde kam (die mancher noch beklaget) da deines Lebens-Uhr den Garaus endlich schlug. Nun ruhest du mit Lust / die Last blieb auf der Erden / Diß Meisterstück' ist nicht zerbrochen / nur zerlegt! Was jetzt vertheilet ist / soll einst ergänzet werden: Wie daß die Unruh dann sich ewig in uns regt?

(Regina Magdalena Limburger, *Die Bertrübte Pegnesis,* poem on the death of Sigmund von Birken ["Floridan"] [Nuremberg, 1684], quoted from E. Mannack, ed., *Die Pegnitzschäfer: Nürnberger Barockdichtung* [Stuttgart, 1968], pp. 72f.)

59. John Webster, *The Duchess of Malfi,* act 3, scene 5, in *The Works,* ed. Alexander Dyce (London, 1859), p. 83a.

60. Bishop Henry King, "Elegie on Gustavus Adolphus" (1633), in *The Poems,* ed. John Sparrow (London, 1925), p. 75.

61. Thus in thy ebony box
Thou dost inclose us, till the day
Put our amendment in our way,
And give new wheels to our disorder'd clocks.

George Herbert, "Even Song" (before 1633), in *The Works,* ed. F. E. Hutchinson (Oxford, 1941), p. 64.

And, according to Pelham, "That food which refresheth the flesh, must be often taken; the body of man requiring to be at sundry times relieved, even as a

Clocke winding up, to keepe it in motion. It fareth not so with the Spirit;" (Sir William Pelham, *Meditations upon the Gospell by Saint John* [London, 1625], p. 98). I owe this quotation to Francis C. Haber, University of Maryland.

62. The relationship between physiological theory and clockwork will be discussed further in chapter 3.

63. Georg Joach[illegible]cus, *De libris revolutionum Copernici narratio prima* (1540), in Edward Rosen, ed., *Three Copernican Treatises* (New York, 1959), pp. 137–38.

64. "We allow every watchmaker so much wisdom as not to put any motion in his instrument, which is superfluous, or may be supplied an easier way: and shall we not think that nature has as much providence as every ordinary mechanic?" (John Wilkins, *A Discourse Concerning a New Planet. . . .* [London, 1640], in *The Mathematical and Philosophical Works* [1802] [London, 1970], 1:239).

65. Granada, *An Excelleny Treatise of Consideration and Prayer,* p. 356.

66. Philippe de Mornay, *A Worke Concerning the trunesse of Christian Religion* (1581), trans. Philip Sidney and Arthur Golding (London, 1617), pp. 95–96.

67. John Norden, *The Labyrinth of Mans Life, or Vertues Delight and Enuies Opposite* (London, 1614), sign. D2 verso.

> 68. . . . wie viel mehr weiser / verständiger und kunstreicher dieser Meister seyn müsse / welcher den Meister des Uhrwercks selbsten gemacht / und ihme den Verstand unnd Geschicklichkeit das Uhrwerck zu machen / eingegeben hat. Ja der das ganze himmlische Firmament unnd Uhrwerck / Sonne / Mon / Planeten / Zeichen / Sternen /etc. durch seine Allweißheit erschaffen hat / und noch biß auff diese Stund und Minuten durch seine Allmacht und Weisheit erhelt / das solches seinen richtigen Lauff und Gang hat / welcher dann niemands anderst ist / dann der HERR. (Geyger, *Horologium politicum,* p. 68)

69. John Robinson, *Essayes; or Observations Divine and Morall,* 2d ed. (London, 1638), pp. 31–32.

70. Ibid., p. 32.

71. See chap. 1, n. 21.

72. For details, see Maurice *Die deutsche Räderuhr,* 1:16.

73. See chap. 1, n. 20 and p. 15.

74. T. H. Svartengren, *Intensifying Similes in English* (Lund, 1918), p. 441.

75. Thomas Nashe, *The Works,* ed. R. B. McKerrow (1589), 1:84; Ben Jonson, *Cynthia's Revels,* 2.1 (1601), in *The Works,* ed. W. Giffard and F. Cunningham, 3 vols. (London, 1904), 1:161a; John Ray, *A Compleat Collection of English Proverbs* (Cambridge, 1670), quoted from Svartengren, *Intensifying Similes,* p. 134; and John Donne, "The Anatomy of the World" (1611), in *The Complete English Poems,* ed. A. J. Smith (London, 1971), p. 274.

76. Thomas Middleton, *Father Hubbard's Tales* (pre-1627), "To the Reader," in *The Works,* ed. A. H. Bullen, 8 vols. (London, 1885–86), 8:54; Idem, *Blurt, Master Constable,* 2.2 (pre-1627), in *The Works,* 1:38; William Shakespeare, *Henry VI,* 1.2 (ca. 1590), in *The Riverside Shakespeare,* ed. G. Blakemore Evans (Boston, 1974), p. 599; Ben Jonson, *Every Man Out of His Humour,* 2.1 (ca.

1600), in *The Works,* 1:79b; and William Cartwright, *The Ordinary,* 1.5 (ca. 1635), in G. B. Evans, *The Plays and Poems of William Cartwright* (Madison, Wis., 1951), p. 287.

77. Maids are Clocks, the greatest Wheel they show,
goes slowest to us, and makes 's hang on
tedious hopes; the lesser, which are concealed,
being often oyl'd with wishes, flee like desires,
and never leave that motion, till the tongue strikes.

Francis Beaumont and John Fletcher, *Wit without Money,* 6.1 (before 1616), in *The Works,* ed. A. Glover and A. R. Waller, 10 vols. (Cambridge, 1905–12), 2:192.

78. Shakespeare, *Love's Labors Lost,* 3.1 (ca. 1595), in *The Riverside Shakespeare,* p. 174.

79. Thomas Dekker and John Webster, *Westward Ho,* 1.1 (1605), in Thomas Dekker, *The Dramatic Works,* 4 vols., ed. F. Bowers (Cambridge, 1955), 2:321.

80. Thomas Middleton, *A Mad World, My Masters,* 4.1 (1608), in *The Works* 3:317.

81. Ben Jonson, *The Silent Woman,* 4.1 (1609), in *The Works* 1:438a.

82. Carlo M. Cipolla, *Clocks and Culture, 1300–1700* (New York and London, 1967), pp. 66–69.

83. Shakespeare's conception of time, for example, was organic and flexible:

> The variable speed of time, dependent entirely on the emotional state of those experiencing it, is a theme which attracts Shakespeare all through his work, from Lucrece's rhetorical maledictions on Tarquin . . . to Troilus's passionate invective when Cressid is torn from him. (Caroline Spurgeon, *Shakespeare's Imagery and What It Tells Us* [Cambridge, 1965], p. 175)

84. Robert Greene, *A Looking Glass for London and England,* 2.2 (1592), in *The Plays and Poems of Robert Greene,* ed. J. C. Collins, 2 vols. (Oxford, 1905), 1:165.

85. The former quote is by John Webster in *The Works,* ed. Dyce (London, 1859), p. 75b, quoted from F. I. Carpenter, *Metaphor and Simile in the Minor Elizabethan Drama* (New York, 1967), p. 88. The latter quotation is by Shakespeare in *Twelfth Night,* 3.1 (1601–2), in *The Riverside Shakespeare,* p. 425.

86. Michael Drayton, *Of His Ladies Not Coming to London,* quoted from Stürzl, *Der Zeitbegriff in der Elisabethanischen Literatur,* p. 108. Similar in content is the poem by Sir John Suckling:

> That none beguiled be by Time's quick flowing,
> Lovers have in their hearts a clock still going;
> For though Time be nimble, his motions
> Are quicker
> And thicker
> Where Love hath his notions.

Herman Fischer, ed., *Englische Barockgedichte,* English and German (Stuttgart, 1971), p. 214. I owe this quotation to Florian Maurice.

87. Shakespeare, *Henry IV, Pt. 1,* 1.2 (1597), in *The Riverside Shakespeare,* p. 849.

88. John Davies of Hereford, *Respice finem* (ca. 1610), in *The Complete Works,* ed. A. B. Grosart, 2 vols. (Edinburgh, 1878), 2:45.

89. Shakespeare, *Richard II,* 5.5 (1595), in *The Riverside Shakespeare,* p. 836.

Chapter 3. The Clockwork Universe

1. Representative modern treatments of the history of the mechanical philosophy are: Anneliese Maier, *Die Mechanisierung des Weltbildes im 17. Jahrhundert,* Forschung zur Geschichte der Philosophie und Pädagogik, Heft 18 (Leipzig, 1938); Marie Boas, "The Establishment of the Mechanical Philosophy," *Osiris* 10 (1952): 412–541; René Dugas, *Mechanics in the Seventeenth Century* (New York, 1958); E. J. Dijksterhius, *The Mechanization of the World Picture,* trans. C. Dikshoorn (Oxford, 1961); and J. E. McGuire, "Boyle's Conception of Nature," *Journal of the History of Ideas* 33 (1972): 523–42.

2. McGuire, "Boyle's Conception of Nature," p. 524.

3. Robert Boyle, *The Christian Virtuoso* (1690), in *The Works,* ed. Thomas Birch, 6 vols. (London, 1772), 5:513.

4. "Desiderant omnes philosophi recentiores physica mechanice explicari," G. W. Leibniz, *Theoria motus concreti* (1671), art. 58, quoted from A. Maier, *Mechanisierung des Weltbildes,* p. 13.

5. G. W. Leibniz, Letter to Remond, 10 January 1714, in *Die philosophischen Schriften,* ed. C. J. Gerhardt (Berlin, 1887), 3:606.

6. Francis Bacon, *De augmentis scientiarum* (Leiden, 1652), book 2, chap. 5, p. 241.

7. Newton is quoted as having said, similarly, "Later philosophers feign hypothesis for explaining all things *mechanically,* and refer other causes to metaphysicks", in Samuel Johnson, *A Dictionary of the English Language,* 2 vols. (Philadelphia, 1819), s.v. "mechanically," no details given.

8. Peter Sternagel, *Die Artes mechanicae im Mittelalter: Begriffs- und Bedeutungsgeschichte bis zum Ende des 13. Jahrhunderts* (Kallmünz, 1966). See also Henry of Monantheuil who, in his dedication to King Henri IV of France in his edition (1599) of the pseudo-Aristotelian *Quaestiones mechanicae,* pleaded: "Be not, I implore you, repelled by this gift with the title of mechanics, as if it were less than noble, lacking in liberal spirit, and not worthy of a ruler and king; do not despise it at first sight." And he went on to argue that God himself is a mechanic. *Aristotelis Mechanica Graeca, emendata, Latina facta, & Commentariis illustrata ab Henrico Monantholio* (Paris, 1599), Epistola dedicatoria.

9. Samuel Johnson, in his *Dictionary of the English Language,* 2 vols. (Philadelphia, 1819), offers three definitions for *mechanick:* "1. constructed by the laws of mechanicks; 2. skilled in mechanicks, bred to manual labor; and 3. mean, servile, of mean occupation." The *Oxford English Dictionary* (Oxford, 1961), vol. 6, offers for the word *mechanical,* among others, the meaning "involving manual labor, even servile, vulgar," plus various illustrative examples.

Other examples can be found in the concordances of representative authors: Christopher Marlowe: "base dunghill villain, and mechanical, I'll have thy head" (Charles Crawford, *The Marlowe Concordance,* 2 vols. [New York, 1964]); Ben Jonson: "whorsen base fellow! a mechanicall serving-man!" (Charles Crawford, *A Ben Jonson Concordance,* 5 vols. [London, 1923]); William Shakespeare: "hang him, mechanical salt-butter rogue" (*Merry Wives of Windsor,* 2.2.278), "base dunghill villain and mechanicall" (*Henry VI, pt. 2,* 1.3,193) (Martin Spevack, *The Harvard Concordance to Shakespeare* [Cambridge, 1973]); and Alexander Pope: "Lord! Sir, a meer Mechanick! strangely low" (Edwin Abbott, *A Concordance of the Works of Alexander Pope* [New York, 1875]). The ambiguity of the word *mechanical* is also discussed at length with much additional material in Christopher Hill, *Change and Continuity in Seventeenth-Century England* (London, 1974), pp. 255–60.

10. Thomas Sydenham, *De Arte Medica* (1669), quoted from Hill, *Change and Continuity,* pp. 255–56.

11. Representative examples are: Nicole Oresme's reference to the Clockmaker God as "that architect who, it is said, has made all things in number, weight, and measure" (see chapter 2 in this volume); Sir William Petty, an outspoken Baconian who made broad use of quantitative empirical data: "For instead of using only comparative and superlative Words, and intellectual Arguments, I have taken the course (as a Specimen of the Political Arithmetick I have long aimed at) to express myself in Terms of *Number, Weight,* or *Measure*" (*Economic Writings,* ed. Charles Henry Hull [Cambridge, 1899], 1:244); and Joseph Addison: "Sir Isaac Newton, who stands up on the Miracle of the Present Age, can look through the whole Planetary System; consider it in its Weight, Number, and Measure; and draw from it as many Demonstrations of infinite Power and Wisdom," (*The Spectator* 543: 211–12, quoted from Michael Schneider, "Of Mystics and Mechanism: The Reception of Newtonian Physics in Eighteenth Century English Poetry," *Synthesis* 1 [1972]: 17).

12. I do not believe that for exciting in us tastes, smells, and sounds there are required in external bodies anything but sizes, shapes, numbers, and slow or fast movements; and I think that if ears, tongues, and noses were taken away, shapes and numbers and motions would remain but not smells or tastes or sounds. (Galileo Galilei, *Il Saggiatore* [1623], question 48, in Galileo Galilei, *Le opere di Galileo Galilei,* ed. Antonio Favaro, 20 vols. [Florence, 1890–1909], 6:350)

> Concerning my method, I thought it not sufficient to use a plain and evident style in what I have to deliver, except I took my beginning from the very matter of civil government, and thence proceeded to its generation and form, and the first beginning of justice. For everything is best understood by its constitutive causes. For as in a watch, or some such small engine, the matter, figure, and motion of the wheels cannot well be known, except it be taken insunder and viewed as parts; so to make a more curious search into the rights of states and duties of subjects, it is necessary, I say not to take them insunder, but yet that they be considered as if they were dissolved. (Thomas Hobbes, *De Cive* [1642], preface, in *The English Works,* ed. Sir William Molesworth, 11 vols. [London, 1842], 2:xiv)

The last sort of Philosophers are such as are wont to be called *Mechanical.* They that list themselves under this Banner, imagine they can explain all the Phenomena of Nature by Matter and Motion, by the Figure and Texture of the Parts, by subtle Particles, and the Actions of Effluvia; and they likewise contend, that these Operations are brought about by the known and establish'd Laws of Mechanics. (John Keill, *An Introduction to Natural Philosophy, or Philosophical Lectures* [London, 1720], lecture 3)

13. Boyle, *About the Excellency and Grounds of the Mechanical Hypothesis* (1665), in *Works,* 4:70.

14. There remains but one course for the recovery of a sound and healthy condition,—namely, that *the entire work of the understanding be commenced afresh,* and that the mind itself be from the very outset not left to take its own course, but guided at every step; and the business be done as if by machinery. (Francis Bacon, *Novum Organum,* Preface, in Bacon, *The Works,* ed. James Spedding, R. L. Ellis, and D. D. Heath, 14 vols. [London, 1857–74, repr. New York, 1968] 1:152)

Whatever quotations such as this may prove about Bacon's status as a mechanical philosopher, the mechanists of the following generation did not hesitate to claim him as one of their own, as for example, A. Cowley did in his ode *To the Royal Society:*

From Words, which are but Pictures to the Thought . . . ,
To Things, the Minds right Object, he [Bacon] it brought,
Like foolish Birds to painted Grapes we flew;
He sought and gather'd for our use the Tru;
And when on heaps the chosen Bundles lay,
He prest them wisely the Mechanic way.

Thomas Sprat, *The History of the Royal Society of London for the Improving of Natural Knowledge* (London, 1667), p. B2.

15. Boyle, *Some Specimens . . . of the Corpuscularian Philosophy,* in *Works,* 1:356.

16. Boyle, *The Excellency of Theology Compared with Natural Philosophy* (1665), in *Works,* 4:49.

17. "Wer nun die Welt als eine solche Maschine ansieht, und alle Begebenheiten und Theile derselben aus der Art ihrer Zusammensetzung und nach den Gesetzen der Bewegung zu erklären sucht; *der philosophiert mechanisch*" (Johann Christoph Gottsched, *Erste Gründe der gesamten Weltweisheit,* 2 vols. [Leipzig, 1733–34], p. 173).

18. "Die Erklärungsart der spezifischen Verschiedenheit der Materien durch die Beschaffenheit und Zusammensetzung ihrer kleinsten Teile, als Maschinen, ist die mechanische Naturphilosophie" (Immanuel Kant, *Atomistik oder Korpuskularphilosophie,* quoted from Rudolf Eisler, *Kant-Lexikon* [Berlin, 1930], p. 348).

19. A well-known example is William Harvey's view of the heart as a pump (see George Basalla, "William Harvey and the Heart as a Pump," *Bulletin of the*

History of Medicine 36 [1962]: 467–70, and Gweneth Whitteridge, *William Harvey and the Circulation of the Blood* [London, 1971], pp. 169–71). Harvey had a wide and subtle interest in mechanical analogies for physiological processes, as is demonstrated in a chapter on "Similitudes and Analogies" in his *De Motu locali animalium,* ed. Gweneth Whitteridge (Cambridge, 1959), pp. 150–51; see also F. N. L. Poynter, "John Donne and William Harvey," *Journal of the History of Medicine* 15 (1960): 242–43.

20. Useful introductions into this subject are P. A. Kocher, *Science and Religion in Elizabethan England* (San Marino, Calif., 1953); Richard S. Westfall, *Science and Religion in Seventeenth-Century England* (New Haven, Conn., 1958); and Reijer Hooykas, *Religion and the Rise of Modern Science* (London, 1972).

21. The "argument from design" is also discussed in this book on pp. 39, 48, 57, and 122f.

22. Hero of Alexandria, *The Pneumatics,* trans. and ed. Bennett Woodcroft (London, 1851); and Vitruvius, *De architectura libri decem,* 9.1.15, ed. Frank Granger, 2 vols. (Cambridge, Mass., 1931), 2:225.

23. All surviving sources on Archimedes' planetarium are quoted in Derek J. Price, "On the Origins of Clockwork, Perpetual Motion Devices, and the Compass," *U.S. National Museum Bulletin 218* (Washington, D.C., 1959), pp. 81–112.

24. Cicero, *De natura deorum,* book 2, chap. 34, quoted by Price in "On the Origins of Clockwork." Eusebius, *Syrische Theophanie,* ed. H. Grossmann, in *Werke,* (Leipzig, 1904), 3.2:68ff.

25. Derek J. Price, *Gears from the Greeks* (New York, 1975); and Otto Neugebauer, "The Early History of the Astrolabe," *Isis* 40 (1949): 240–55.

26. For descriptions of examples, see K. Maurice and O. Mayr, *The Clockwork Universe: German Clocks and Automata 1550–1650* (Washington, D.C. and New York, 1980), pp. 290–301.

27. For examples of early astronomical clocks with heliocentric indications, see ibid., pp. 222–23 and 304–6.

28. David Pingree, "Tycho Brahe," *Dictionary of Scientific Biography* (New York, 1970), 2:406–9.

> 29. Annon ridiculum esset comparare totum illud corporis spatiosissimi systema, ad Saturni sideris corpus? annon multis millium millibus ipsum superaret? ac si quis rotam fingeret, cujus diameter esset unium milliaris, crassities vero compaginis triens ejusdem? rotam vero hanc factam, ut pilam palmariam volutaret. (*De mundo,* 154, in Suzanne Kelly, *The De Mundo of William Gilbert* [Amsterdam, 1965], p. 154)

30. Francis Bacon, *De augmentis scientiarum,* book 3, chap. 4, in *The Works,* ed. Spedding, Ellis, and Heath, 4:348–49.

> 31. Scopus meus hic est, ut coelestem machinam dicam non esse instar divini animalis, sed instar horologii (qui horologium credit esse animatum, is gloriam artificis tribuit operi), ut in qua pene omnis motuum varietas ab una simplicissima vi magnetica corporali, uti in horologio motus omnes a sim-

plicissimo pondere. Et doceo hanc rationem physicam sub numeros et geometriam vocare, ne quid mihi metuas a somniis Alpetragii, qui ante Fracastorium omnia conatus est concentricis efficere: sed numeros non attigit, alias adhaesisset et se somniari intellexisset. (Johannes Kepler, Letter to Herwart von Hohenburg, Prague, 10 February 1605, in *Opera Omnia,* ed. C. Frisch [Frankfurt, 1859], 2:83–84)

32. Hooykaas, *Religion and the Rise of Modern Science,* p. 62.

33. This issue frequently surfaces in the vast and diffuse literature on English literary history. For a good survey, see Peter H. Niebyl, "Science and Metaphor in the Medicine of Restoration England," *Bulletin of the History of Medicine* 47 (1973): 356–74.

34. Sprat, *History of the Royal Society,* pp. 39 and 42.

35. Ibid., p. 327 and pp. 112–13.

36. Niebyl offers many examples; see n. 33 above, passim. Boyle, who loved metaphors, frequently apologized for this aberration, and John Locke expressed himself against "figurative speech" and "ornaments" in philosophy (*An Essay concerning Human Understanding,* 3d book in John Locke, *The Works,* ed. Edmund Law, 4 vols. [London, 1777, repr. 1824], 2:288ff.). Hume, in refuting the argument from design, argued against proof by analogy (David Hume, *Dialogues concerning Natural Religion,* ed. Norman Kemp Smith [New York, 1948], pp. 141–57).

The playwright Thomas Shadwell called one of the villains of his play *The Virtuoso* (1676) Sir Formal Trifle, "the greatest Master of Tropes and Figures" (Act 1, *The Complete Works,* ed. M. Summers [London, 1927], quoted from Walter E. Houghton, "The English Virtuoso in the Seventeenth Century," *Journal of the History of Ideas* 3 [1942]: 199).

37. "La grande Mechanique n'étant autre chose que l'ordre que Dieu a imprimé sur la face de son ouvrage, que nous appelons communement la Nature," (René Descartes, Letter to Villebressieu, Amsterdam, Summer 1631, in C. Adam and P. Tannery, eds., *Oeuvres de Descartes,* 13 vols. [Paris, 1898–1913], 1:213–14); and "Tertium denique quod notandum duxi, est versus finem in his verbis: *materia mundi machina existit.* Ubi potius scripsissem: *mundum machinae instar ex materia constare,* vel *esse easdem omnes in rebus materialibus motuum causas, atque in mechinis arti factis,* vel quid simile" (Descartes, Letter to unknown, ca. 1640, in *Oeuvres,* 5:546).

38. Descartes, *Principia philosophiae,* pt. 4, chap. 188, in *Oeuvres,* 8A:315.

39. "Ita consistit universum totum in aequi pondio. Sed hoc difficillimum conceptum est, quia mathematicum et mechanicum est; nos autem machinas non satis assueti sumus considerare, et hinc omnis fere error in philosophiâ exorsus est" (Descartes, Interview with Frans Burman, Egmond, 16 April 1648, in *Oeuvres,* 5:174).

40. . . . ie les veux avertir que ce mouvement, [the functioning of the heart] que ie vien d'expliquer, suit aussy necessairement de la seule disposition des organes qu'on peut voir a l'œil dans le cœur, & de la chaleur qu'on y peut sentir avec les doigts, & de la nature du sang qu'on peut connoistre par

experience, que fait celuy d'un horologe, de la force, de la situation, & de la figure de ses contrepois & de ses rouës. (Descartes, *Discours de la méthode,* pt. 5, in *Oeuvres,* 6:50)

41. Descartes, *Principia philosophiae,* pt. 4, chap. 203, in *Oeuvres,* 8A:326.

42. Ibid., pt. 4, chap. 204, in *Oeuvres,* 8A:327.

43. For general background, see G. Canguilhem, "Descartes et la technique," in *Congrès Descartes: Traveaux du IXe Congrès International de Philosophie* (Paris, 1937), 2:77–85; J. Baltrusaitis, "Descartes: Les Automates et le doute," in *Anamorphoses ou Perspectives Curieuses* (Paris, 1955), pp. 33–43; and G. Rodis-Lewis, "Machineries et Perspectives curieuses dans leurs rapports avec le Cartésianisme," *Le Dix-Septième Siècle* 6.32 (1956): 461–74.

44. References to characteristic passages, especially on hydraulic machinery and its historical background can be found in Descartes, *Oeuvres,* 11:212–15, 669; and Charles Adam, *Vie et oeuvres de Descartes: Étude historique, Oeuvres, 12:163, note c.*

45. Arthur H. Schrynemakers, "Descartes and the Weight-Driven Chain Clock," *Isis* 60 (1969): 233–36.

46. Descartes's definition of an automaton ("automates, ou machines mouvantes," Descartes, *Discours de la méthode,* in *Oeuvres,* 6:55) was that suggested by the word's etymology: When he spoke of "a clock, or some other automaton" ("horologium, aliudve automatum," Descartes, Letter to Regius, Endegeest, June 1642, in *Oeuvres,* 3:566), or again of "a watch, or some other automaton—that is, some other machine that moves of itself" ("une montre, ou autre automate—c'est à dire, autre machine qui se meut de soy-mesme," Descartes, *Les Passions de l'Ame,* pt. 1, art. 6, in *Oeuvres,* 11:331), he was simply following contemporary usage, which held that automata and clocks were the same thing or, more precisely, that clocks were a special class within the larger category of automata (see also pp. 21–26 and n. 23, chap. 1 above). Throughout his writing he tended to employ the terms *automaton, clock,* and *self-moving machine* interchangeably.

47. Otto Mayr, "Automatenlegenden in der Spätrenaissance," *Technikgeschichte* 41 (1974): 20–32.

48. Concerning Descartes's knowledge of these technological showpieces, see n. 44, above).

49. Descartes, *Cogitationes privatae,* in *Oeuvres,* 10:231–32.

50. "Je ne reconnois aucune difference entre les *machines que font les artisans* & les divers corps *que la nature seule compose*" (Descartes, *Principia philosophiae,* pt. 4, chap. 203, in *Oeuvres,* 9B:321).

51. "Non sit tam probabile omnes vermes, culices, erucas, & reliqua animalia immortali animâ praedita esse, quam machinarum instar se movere" (Descartes, Letter to Henry More, Egmond, 5 February 1649, in *Oeuvres,* 5:277).

52. Descartes, *Discourse on Method,* trans. Paul J. Olscamp (Indianapolis, 1965), p. 45, also in *Oeuvres,* 6:56.

53. Mais si nous estions aussi accoustumez à voir des automates, qui imitassent parfaitement toutes celles de nos actions qu'ils peuvent imiter, & à ne les prendre que pour des automates, nous ne douterions aucunement que tous

les animaux sans raison ne fussent aussi des automates, à cause que nous trouverions qu'ils different de nous en toutes les mesmes choses, comme i'ay écrit page 56 de la Methode. Et i'ay deduit tres-particulierement en mon Monde, comment tous les organes qui sont requis à un automate, pour imiter toutes celles de nos actions qui nous sont communes avec les bestes, se trouvent dans le cors des animaux. (Descartes, Letter to Mersenne, 30 July 1640, in *Oeuvres,* 3:121)

His detailed references to earlier formulations underscore the importance he assigned to the argument. He also extended this notion to humans:

d'où ie voudrois presque conclure, que l'on connoist la cire par la vision des yeux, & non par la seule inspection de l'esprit, si par hazard ie ne regardois d'une fenestre des hommes qui passent dans la ruë, à la veuë desquels ie ne manque pas de dire que ie voy des hommes, tout de mesme que ie dis que ie voy de la cire; Et cependant que voy-je de cette fenestre, sinon des chapeaux & des manteaux, qui peuvent couvrir des spectres ou des hommes feints qui ne se remuent que par ressors? (Descartes, *Meditation No. 2,* in *Oeuvres,* 9:25)

Enfin il n'y a aucune de nos actions exterieures, qui puisse assurer ceux qui les examinent, que nostre cors n'est pas seulement une machine qui se remuë de soy-mesme, mais qu'il y a aussi en luy une ame qui a des pensées, excepté les paroles, ou autres signes faits à propos des suiets qui se presentent, sans se raporter à aucune passion. (Descartes, Letter to the Marquis of Newcastle, Egmond, 23 November 1646, in *Oeuvres,* 4:574)

54. Descartes, *Oeuvres,* 11:119–290. Critical editions of this work have recently appeared under the titles, *Treatise of Man,* trans. and ed. T. S. Hall, Harvard Monographs in the History of Science, No. 6 (Cambridge, 1972), and *Über den Menschen, sowie Beschreibung des menschlichen Körpers,* trans. and ed. K. E. Rothschuh (Heidelberg, 1969).

55. For example, near the beginning of the *Traité de l'Homme* (Hall, *Treatise of Man,* pp. 1–5), or later in Olscamp, *Discourse on Method,* pp. 44–45.

56. See n. 53, above, and also *Discourse de la Méthode,* in *Oeuvres,* 6:53–55.

57. "A sçavoir, mon opinion est que cete glande est le principal siege de l'ame, & le lieu ou se font toutes nos pensées" (Descartes, Letter to Meyssonier, 29 January 1640, in *Oeuvres,* 3:19). See also *Traité de l'Homme,* passim.

58. Descartes, in Hall, *Treatise of Man,* p. 91.

59. Norbert Wiener, *Cybernetics, or Control and Communication in the Animal and the Machine* (Cambridge, Mass., 1948).

60. Hall, *Treatise of Man,* p. 34.

61. Descartes, however, was not incapable of perceiving more complex communication schemes connecting various systems of organs: in the action of a hand grasping some object, for example, he recognized some form of cooperation between eyes, brain, and hand that required reciprocal exchanges of information between the participating organs. (See Rothschuh, *Über den Menschen,* p. 122, and Hall, *Treatise of Man,* pp. 31 and 98.)

His account of this process has been interpreted by modern readers as an attempt to describe a closed feedback loop. One hesitates to accept this interpretation because here, for once, Descartes did not make himself sufficiently clear, perhaps for lack of a tangible mechanical model. At any rate, he failed to articulate the concept of the closed loop well enough that his contemporaries could comprehend it and put it to active use. He himself, too, overlooked the possibilities that this concept would have opened up for physiological theory: the concept of a cooperative interaction between the command function of the brain and the perceptions of the senses would have made it possible to visualize systems that were able to react appropriately to unrehearsed situations and to meet challenges for which the correct response would not be programmed in advance. Such systems would no longer be controlled by a rigid program; instead they would have been free, within certain limits, to act according to decisions of their own. These possibilities Descartes did not exploit. His physiological thought remained under the domination of the analogy of the program-controlled automaton.

62. Descartes, *Discours de la Méthode,* in *Oeuvres,* 6:57, or *Meditation No. 6,* in *Oeuvres,* 9:A67; and Letter to the Marquis of Newcastle in *Oeuvres,* 4:575.

> 63. Et le second est que, bien qu'elles [machines] fissent plusieurs choses aussy bien, ou peutestre mieux qu'aucun de nous, elles manqueroient infailliblement en quelques autres, par lesquelles on découvriroit qu'elles n'agiroient pas par connoissance, mais seulement par la disposition de leurs organes. Car, au lieu que la raison est un instrument universel, qui peut servir en toutes sortes de rencontres, ces organes ont besoin de quelque particuliere disposition pour chaque action particuliere; d'où vient qu'il est moralement impossible qu'il y en ait assez de divers en une machine, pour la faire agir en toutes les occurrences de la vie, de mesme façon que nostre raison nous fait agir. (Descartes, *Discours,* in *Oeuvres,* 6:57)

> 64. Car on peut bien concevoir qu'une machine soit tellement faite qu'elle profere des paroles, & mesme qu'elle en profere quelques unes a propos des actions corporelles qui causeront quelque changement en ses organes: comme, si on la touche en quelque endroit, qu'elle demande ce qu'on luy veut dire; si en un autre, qu'elle crie qu'on luy fait mal, & choses semblables; mais non pas qu'elle les arrenge diversement, pour respondre au sens de tout ce qui se dira en sa presence, ainsi que les hommes les plus hebetez peuvent faire. (Descartes, *Discours,* in *Oeuvres,* 6:56–57)

65. Descartes, Letter to Reneri, April or May 1638, in *Oeuvres,* 2:40.

66. Descartes expressed this evaluation many times throughout his life, for example:

—in 1619: Ex animalium quibusdam actionibus valde perfectis, suspicamur ea liberum arbitrium non habere (Descartes, *Cogitationes privatae,* in *Oeuvres,* 10:219);

—in 1637: [the fact that animals can do certain things] better than we, does not prove that they have minds.

> De façon que ce qu'ils font mieux que nous, ne prouve pas qu'ils ont de l'esprit; car, a ce conte, ils en auroient plus qu'aucun de nous, & feroient mieux en toute chose; mais plutost qu'ils n'en ont point, & que c'est la Nature qui agist en eux, selon la disposition de leurs organes: ainsi qu'on voit qu'un horologe, qui n'est composé que de rouës & de ressors, peut conter les heures, & mesurer le tems, plus iustement que nous avec toute nostre prudence. (Descartes, *Discours,* in *Oeuvres,* 6:58–59)

—or in 1646:

> Ie sçay bien que les bestes font beaucoup de choses mieux que nous, mais ie ne m'en estonne pas; car cela mesme sert à prouver qu'elles agissent naturellement & par ressors, ainsi qu'une horloge, laquelle monstre bien mieux l'heure qu'il est, que nostre jugement ne nous l'enseigne. Et sans doute que, lors que les hirondelles viennent au printems, elles agissent en cela comme des horloges. (Descartes, Letter to the Marquis of Newcastle, in *Oeuvres,* 4:575)

67. Blaise Pascal, *Pensées,* ed. and trans. H. F. Stewart (New York, 1950), nos. 286 and 285. Subsequent references are to this edition.

68. H. J. M. Bos, "Christiaan Huygens," *Dictionary of Scientific Biography* (New York, 1972), 6:597–613.

69. Christiaan Huygens, *Oeuvres complètes,* 22 vols. (The Hague, 1888–1950), 10:400.

70. Baruch de Spinoza, *Tractatus de intellectus emendatione,* in *Opera,* ed. Carl Gebhardt, 4 vols. (Heidelberg, 1925), 2:18.

71. Ibid., p. 32. This remark was quoted with interest and approval by G. W. Leibniz in *The Monadology and Other Philosophical Writings,* trans. and ed. Robert Latta (Oxford, 1925), p. 230.

72. Nicolas Malebranche, *Treatise concerning the Search after Truth,* trans. T. Taylor (London, 1700), 1:51.

73. Ibid., 1:162.

74. I grant that all the actions that Beasts perform are certain indications of an Intelligence; for whatever is regular demonstrates it. A Watch shews the same; for 'tis impossible Chance should have composed its Wheels, but an understanding Agent must have ordered its Motions. . . .

> The Motions of Beasts and Plants intimate an Intelligence, but that Intelligence is not Matter, and is much distinguished from Beasts, as that which disposes the Wheels of a Watch is distinguished from the Watch itself. (Ibid., 2:77)

75. "We easily see that a Watch points out the Hour, but it requires time to find out the Reasons of it. And there are so many different Movements in the Brain of the least of Animals, as far surpass the most compound Clock-work of the World" (ibid., 1:166).

"And if the Idea of a Watch which shows the Hour, with all the different Motions of the Planets, were no perfecter than that of another, which only points to the hour, or than that of a Circle and a Square, a Watch would be no perfecter than a Circle" (ibid., 2:136).

76. "The same *subtil matter* which causes . . . [the effects of animation] in a Dog, and which is the principle of his Life, is no perfecter than that which gives Motion to the Spring of a Watch, or which causes the Gravitation in the Weights of a Clock" (ibid., 2:157).

77. "The *Soul* may be taken for something *Corporeal* . . . ; or else for something *Spiritual.* Those that pretend that Beasts have no Soul, understand it in the second Sense; for never any Man denied that there is in Animals something *Corporeal,* which is the Principle of their Life or Motion; since it cannot be denied even of Watches" (ibid., 2:76).

78. "Yet there are Men so little capable of Reflexion, that we might almost compare them with *Machines* purely inanimate" (ibid., 1:166).

> 79. For even as to our being perswaded that a *Watch* hath no Entity, different from the Matter, it is compos'd of; it suffices to know, how the different Disposition of the Wheels is able to effect all the Movements of a *Watch,* without having any other distinct Idea of what might possibly be the Cause of these Motions, though there be many *Logical* to [be] had. So because these Persons have no distinct Idea of what could be in Matter, were Extension taken away, and see no Attribute that can explicate its Nature, and because Extension being granted, all the Attributes conceiv'd to belong to Matter, are at the same time granted: and because Matter is the Cause of no Effect, which may not be conceiv'd producible by Extension, diversly configur'd, and diversly mov'd; therefore they are perswaded, that Extension is the Essence of Matter.
>
> But as no Man can infallibly demonstrate there is not some Intelligence, or New-created Entity in the Wheels of a *Watch;* so no man can, without particular Revelation, be assur'd as of a *Geometrical* Demonstration, that there is nothing but Extension diversely configured in a Stone. For 'tis absolutely possible for Extension to be joyn'd with something which we don't conceive, because we have no Idea of it: though it seems very unreasonable to believe and assert it; it being contrary to Reason, to assert what we neither know, nor have any conception of. (Ibid., 1:126)

80. Fénélon, François de Salignae de la Mothe, *A Demonstration of the Existence and Attributes of God* (1720), trans. A. Boyer (Harrisburg, Pa., 1811).

81. "Now, I ask what mover gave motion to that first atom, and first set the great machine of the Universe going?" (Ibid., p. 130)

"The more the great spring, that directs the machine of the Universe is exact, simple, constant, certain, and productive of abundance of useful effects; the more it is plain, that a most potent, and most artful hand knew how to pitch upon the spring which is the most perfect of all" (ibid., p. 32).

"Matter alone cannot either by simple laws of motion, or by the capricious strokes of Chance, make even animals that are mere machines. Those philosophers themselves who will not allow beasts to have a reasoning faculty, cannot avoid acknowledging that what they suppose to be blind and artless in these machines, is yet full of wisdom and art in the first mover, who made their springs and regulated their movement" (ibid., p. 17).

> 82. It may be said, that the motion of the stars is settled and regulated by unchangeable laws. . . . Whence proceeds the government of that universal machine, which incessantly works for us, without so much as our thinking upon it? To whom shall we ascribe the *choice* and gathering of so many deep, and so well concerted springs; and of so many bodies, great and small, visible and invisible, which equally concur to serve us? The least atom of this machine, that should happen to get out of order, would unhinge all nature. For the springs and movements of a watch are not put together with so much art and niceness, *as those of the Universe.* What then must be a design so extensive, so coherent, so excellent, so beneficial? The necessities of those laws, instead of deterring me from inquiring into their author, does but heighten my curiosity, and admiration. (Ibid., p. 34)

83. "The command of my mind over my body is supreme, and absolute in its bounded extent, since my single will, without any effort, or preparation, causes all the members of my body, to move on a sudden, and immediately, according to the rules of mechanicks" (ibid., p. 77).
"But the soul that governs the machine of man's body, moves all its springs in time, without seeing or discerning them; without being acquainted with their figure, situation, or strength, and yet it never mistakes" (ibid., p. 79).

> 84. We cannot sufficiently admire either the absolute power of the soul over the corporeal organs which she knows not, or the continual use it makes of them discerning them. That sovereignty principally appears with respect to the images imprinted in our brain. I know all the bodies of the Universe that have made any impression on my senses for a great many years past. I have distinct images of them. . . . My brain is like a closet of pictures. . . . Moreover, all these images either appear, or retire as I please, without any confusion: I call them back and they return. I dismiss them, and they sink I know not where. (Ibid., pp. 74–75)

85. Some representative titles are the following: George Boas, *The Happy Beast in French Thought of the Seventeenth Century* (Baltimore, 1933); Leonora D. Cohen, "Descartes and Henry More on the Beast Machine: A Translation of their Correspondence Pertaining to Animal Automation," *Annals of Science* 1 (1936): 48–61; Leonora Cohen Rosenfield, "Un Chapitre de l'histoire de l'animal machine (1645–1749)," *Revue de Littérature Comparée* 17 (1937): 461–87; Leonora Cohen Rosenfield, *From Beast-Machine to Man-Machine: Animal Soul in French Letters from Descartes to La Mettrie* (New York, 1941); Aram Vartanian, *Diderot and Descartes: A Study of Scientific Naturalism in the Enlightenment* (Princeton, 1960); Heikki Kirkinen, *Les Origines de la conception moderne de l'homme-machine, 1670–1715* (Helsinki, 1960); Wallace Shugg, "The Cartesian Beast Machine in English Literature (1663–1750)," *Journal of the History of Ideas* 29 (1968): 279–92; Thomas S. Hall, "Mechanism, Machines and Mechanistic Biology: An Introduction," in *Ideas of Life and Matter,* 2 vols.

(Chicago, 1969), 1:218–29; William P. D. Wightman, "Myth and Method in Seventeenth Century Biological Thought," *Journal of the History of Biology* 2 (1969): 321–36; Blair Campbell, "La Mettrie: The Robot and the Automaton," *Journal of the History of Ideas* 31 (1970): 555–72; Julian Jaynes, "The Problem of Animate Motion in the Seventeenth Century," *Journal of the History of Ideas* 31 (1970): 219–34; and Thomas S. Hall, "Descartes' Physiological Method: Position, Principles, Examples," *Journal of the History of Biology* 3 (1970): 53–79.

86. Bernard le Bovier de Fontenelle, *A Plurality of Worlds,* trans. John J. Glanvell (London, 1688), in Leonard M. Marsak, ed. and trans., *The Achievement of Bernard le Bovier de Fontenelle* (New York and London, 1970), pp. 10–11.

87. See chap. 3, n. 5, above.

88. E. Gerland, ed., *Leibnizens nachgelassene Schriften physikalischen, mechanischen und technischen Inhalts* (Leipzig, 1906).

89. For a good historical survey of the argument, see Leibniz, *Monadology,* introduction, esp. pp. 42–47.

90. Leibniz, Letter to Arnauld, 1687, in *Monadology,* p. 47.

91. Leibniz, "Third Explanation," *Journal des Savants* (November 1696), in *Monadology,* pp. 331–34. Leibniz had published his system of preestablished harmony first in an article, "New System of the Nature of Substances and of the Communication between Them, as Well as of the Union There Is between Soul and Body," *Journal des Savants* (June 1695). In the ensuing discussion, he offered three further "Explanations."

92. Leibniz, *Monadology,* pp. 314 and 315.

93. Mechanism is sufficient to produce the organic bodies of animals . . . provided there be added thereto the *preformation* . . . in the seeds of the bodies. . . . This could only proceed from the Author of things, infinitely powerful and infinitely wise, who . . . had *pre-established* there all order and artifice that was to be. . . . In order to explain this marvel of the formation of animals, I made use of a Pre-established Harmony, that is to say, of the same means I had used to explain another marvel, namely the correspondence of soul and body. (G. W. Leibniz, *Theodicy* [1710], ed. Austin Farrar, trans. E. M. Huggard [New Haven, 1952], pp. 64–65)

94. Ibid., p. 65.

95. Ibid., p. 151.

96. Ibid., p. 157.

97. The foetus forms itself in the animal, and a thousand other wonders of nature are produced by a certain *instinct* that God has placed there, that is by virtue of *divine preformation,* which has made these admirable automata, adapted to produce mechanically such beautiful effects. Even so it is easy to believe that the soul is a spiritual automaton still more admirable, and that it is through divine preformation that it produces those beautiful ideas, wherein our will has no part and to which our art cannot attain. The operation of spiritual automata, that is of souls, is not mechanical, but it contains in the highest degree all that is beautiful in mechanism. (Ibid., pp. 364–65)

98. "The organic body of each living being is a kind of divine machine or natural automaton which infinitely surpasses all artificial automata. . . . The machines of nature, namely, living bodies, are still machines in their smallest parts *ad infinitum*" (*Monadology*, p. 254).

99. "All simple substances or created Monads might be called Entelechies, for they have in them a certain perfection . . . they also have in them a certain self-sufficiency (αὐτάρκεια) which makes them the sources of their internal activities and, so to speak, incorporeal automata" (ibid., p. 229).

100. H. G. Alexander, ed. *The Leibniz-Clarke Correspondence* (Manchester, 1956).

101. Christian Wolff, *Vernünfftige Gedancken von Gott, der Welt und der Seele des Menschen* (1719) (Halle, 1747).

102. "So folgt . . . , dass die Welt eine Reihe unveränderlicher Dinge sey, die neben einander sind, und auf einander folgen, insgesamt aber mit einander verknüpffet sind. . . . so ist die Welt als ein Ding anzusehen" (ibid., §§332 and 333).

103. "Und solchergestalt sind die Dinge in der Welt der Zeit nach verknüpffet, weil sie dem Raume nach miteinander verknüpffet sind" (ibid., §554).

104. "Es verhält sich die Welt nicht anders als wie ein Uhrwerck. Denn das Wesen der Welt bestehet in der Art ihrer Zusammensetzung; das Wesen einer Uhr gleichfalls" (ibid., §556). Subsequent § references in the text are to Wolff.

105. "Man siehet leicht, dass, was von der Welt gesaget worden, auch von allen zusammengesetzten Dingen gilt: nehmlich, dass auch sie Maschinen sind und eben deswegen in ihnen Wahrheit ist" (ibid., §560).

106. For example, in §§565, 566, 568, 569, 578, 612, 639, 640, 714, and 716.

107. And for this I take a clock that can run for a whole year and that has been wound up and set no more than once. In this case nobody will doubt that the movements of the hand and of every wheel have their cause in the type of design, and hence in the composition and movements of all the other pieces that are properly part of the clock; in other words the hand would not be so pointed at the present instant, if it had not been pointed the way it actually has at all conceivable moments in the time that has elapsed since the start. In the clock this has no other reason than that [its] movements have their sufficient reason in [its] particular type of design. Since this also applies to the world, it must be true that no event, no matter how small, would ever happen if not all preceding events had taken place in the manner determined by earlier times, and if not the space of the world was filled in the manner in which we find it filled. (Ibid., §565)

108. "Here it would be again the same as in our simile of the clock. File off a little from the smallest tooth of a wheel; and the clock will forever run differently from the way it would have run if nothing had been filed off, and indeed it is no longer the same clock" (ibid., §568).

109. So that it can be better visualized what a miracle is I will illustrate it with a simile. I have shown above that the happenings of the world have the very same make-up as the motions of a clock. The orderly motions that result from

> the particular design, with the aid of the force that moves it, are like natural happenings: if, however, the hand is pushed from outside, then this motion is not the result of the composition of parts and the propulsive force of the clock, and thus resembles a miracle in nature. (Ibid., §637)

110. Ibid., "Das erste Register," s.v. Zufällig.

111. Leonhard Euler, *Letters . . . on Different Subjects in Natural Philosophy Addressed to a German Princess,* ed. David Brewster, 2 vols. (New York, 1838), letter 84, 1:280.

112. Christian Wolff, *Cosmologia Generalis* (Frankfurt and Leipzig, 1737).

113. Ibid., §117, p. 107.

114. The clockwork metaphor has a similar and equally prominent role in the book of Wolff's disciple, Johann Christoph Gottsched, *Erste Gründe der gesamten Weltweisheit,* 2 vols. (Leipzig, 1733–34), 1:172–75.

> 115. When I see a watch whose hand marks the hours, I conclude that an intelligent being has arranged the springs of that machine with the purpose that the hand indicate the hours. Likewise, when I see the springs of the human body, I conclude that an intelligent being has arranged those organs so that they should be conceived and nourished for nine months in the womb; that the eyes are for seeing, and the hands for grasping, etc. (Voltaire, *Traité de métaphysique* (1734), ed. H. Temple Patterson [Manchester, 1937], pp. 7–8)

116. Voltaire, *Eléments de la philosophie de Newton,* in Voltaire, *Oeuvres complètes,* ed. Louis Moland, 52 vols. (Paris, 1883–85), 22:404, 405–6.

117. "When we see a fine machine, we say there is a good machinist, and that he has an excellent understanding. The world is assuredly an admirable machine; therefore there is in the world, somewhere or other, an admirable intelligence. This argument is old but is not therefore the worse" (Voltaire, *A Philosophical Dictionary,* 6 vols. [London, 1824], 1:314, s.v. Atheism).

In "Les Cabales" he stated that "the universe embarrasses me, and I cannot imagine that this clockwork exists and has no clockmaker" (Voltaire, *Oeuvres complètes,* ed. Louis Moland, 52 vols. [Paris, 1883–85] 10:182).

118. Voltaire, Letter to Frederick II of Prussia, Cirey, October 1737, in *Oeuvres complètes* (Paris, 1880), 34:321.

> 119. "How do we know what our soul is, we who can form no idea whatever of the [nature of] light when we had the misfortune to be born blind? For I can painfully see that all that has been written about the soul cannot teach us the slightest truth. My principle aim, after having groped about that soul in order to guess its nature, is at least to try to direct it: it is the spring of our clockwork. All of Descartes's fine ideas on elasticity teach me nothing about the nature of this spring; I still don't know the cause of elasticity; meanwhile I wind the clock; it runs just as well. (Ibid.)
>
> The nature of the soul [in contrast to its immorality] is a different matter; religion cares little what the soul is made of so long as it is virtuous. It is a clock we have been given to regulate; but the craftsman who made it has not told us

the composition of the spring. (Voltaire, *Lettres philosophiques,* "Lettre 13 sur M. Locke," Amsterdam, 1734, in *Oeuvres,* ed. Garnier, 22:124–25)

120. On Diderot's attitudes toward freedom and determinism, see, for example, Lester G. Crocker, *Diderot's Chaotic Order: Approach to Synthesis* (Princeton, 1974), pp. 37–39; and John Robert Loy, *Diderot's Determined Fatalist: A Critical Appreciation of Jacques le fataliste* (New York, 1950), pp. 128–60, esp. 134–35.

121. Denis Diderot, *Pensées philosophiques,* in *Oeuvres complètes,* ed. J. Assézat and M. Tourneux, 20 vols. (Paris, 1875–77), 1:132–33, quoted from Lester G. Crocker, ed., *Diderot's Selected Writings,* trans. Derek Coltman (New York, 1966), pp. 3–4.

122. Denis Diderot and Jean d'Alembert, *L'Encyclopédie,* 17 vols. text plus 11 vols. plates (Paris, 1751–66), 10:126.

123. Ibid., 10:220, s.v. Méchanicien.

124. Diderot, *Lettre sur les sourds et muets* (1751), in *Oeuvres complètes,* 1:367.

125. Diderot, *Eléments de physiologie* (1774–80), in *Oeuvres complètes,* 9:265–66.

126. Aram Vartanian, ed., *Julien Offray de la Mettrie, L'Homme machine: A Study in the Origins of an Idea* (Princeton, 1960), p. 190. Subsequent page numbers in the text are to this work.

127. Diderot and d'Alembert, *Encyclopédie,* 15:777, s.v. Système.

128. Ibid., 8:50, s.v. Harmonie.

129. The following is a sampling of representative literature: Marjorie Nicolson, "The Early Stages of Cartesianism in England," *Studies in Philology* 26 (1929): 356–74; Joshua C. Gregory, "Cudworth and Descartes," *Philosophy* 8 (1933): 454–67; Sterling P. Lamprecht, "The Role of Descartes in Seventeenth-Century England," *Studies in the History of Ideas* 3 (1935): 181–240; Cohen, "Descartes and Henry More on the Beast Machine"; J. E. Saveson, "Descartes' Influence on John Smith, Cambridge Platonist," *Journal of the History of Ideas* 20 (1959): 258–63; J. E. Saveson, "Differing Reactions to Descartes among the Cambridge Platonists," *Journal of the History of Ideas* 21 (1960): 560–67; Lydia Gysi, *Platonism and Cartesianism in the Philosophy of Ralph Cudworth* (Bern, 1962); Danton Sailor, "Cudworth and Descartes," *Journal of the History of Ideas* 23 (1962): 133–40; Laurens Laudan, "The Clock Metaphor and Probabilism: The Impact of Descartes on English Methodological Thought, 1650–65," *Annals of Science* 22 (1966): 73–104; Charles Webster, "Henry More and Descartes: Some New Sources," *British Journal for the History of Science* 4 (1969): 359–77; and Shugg, "The Cartesian Beast Machine in English Literature."

130. Robert Boyle, Manuscript Fragment (undated), *Boyle Papers,* Royal Society (London) 2:141.

131. Robert Boyle, *The Works,* 2:7, 39, 45, 48; and 5:163; and Boyle, *Occasional Reflections* (1665), in *Works,* 2:374–76.

132. Boyle, *The Christian Virtuoso,* in *Works,* 5:511.

133. Boyle, *The Usefulness of Natural Philosophy* (1650?), in *Works,* 2:39.

134. Boyle, *The Christian Virtuoso,* in *Works,* 5:511–12.

135. Simon Patrick, *A Brief Account of the New Sect of Latitude-Men* (London, 1662), p. 18.

136. Henry Power, *Experimental Philosophy* (London, 1664), p. 193.

137. Matthew Hale, *The Primitive Origination of Mankind Considered and Examined according to the Light of Nature* (London, 1677), pp. 340–42.

138. Boyle, *A Free Inquiry into the Vulgarly Received Notion of Nature* (1686), in *Works,* 5:245. The same point is made in Boyle, *The Excellency of Theology Compared with Natural Philosophy,* in *Works,* 4:72; and in Abraham Cowley, *College of Prophets,* quoted in Nell Eurich, *Science in Utopia: A Mighty Design* (Cambridge, Mass., 1967), p. 184.

139. Joseph Glanvill, *Scepsis scientifica* (London, 1665), p. 13.

140. Ibid., p. 133.

141. Joseph Glanvill, *Plus ultra: or, The progress and advancement of knowledge since the days of Aristotle* (London, 1668), p. 11.

142. In mechanical things, when any one would observe the motions of a Clock or Engine, he takes the Machine itself to pieces to consider the singular artifice, and doth not doubt but he will learn the causes and properties of the Phaenomenon, if not all, at least the chief: In like manner, when it is brought before your eyes to behold and consider the structure and parts of a muscle, the conformations of the moving fibres, . . . why is it that we should despair to extricate the means or reasons of the motive function . . . ? (Thomas Willis, *Practice of Physick,* 2 vols. [London, 1684], 2:32)

143. Roger Cotes, preface to the second edition of Newton's *Principia* (1713), trans. Andrew Motte (1729), ed. Florian Cajori (Berkeley, Calif., 1947), p. 28, quoted from Laurens Laudan, "The Clock Metaphor and Probabilism," pp. 102–3.

144. Sprat, *The History of the Royal Society,* pp. 82–83.

145. John Locke, *An Essay concerning Human Understanding,* ed. A. C. Fraser (1894), 2 vols. (New York, 1959), 2:216.

146. Robert Hooke, *Micrographia* (London, 1665), preface, no page.

147. Boyle, *The Usefulness of Natural Philosophy* (1650?), in *Works,* 2:45–46; several decades later, in his last book, Boyle still insisted upon this point:

> Suppose a man . . . looks upon a curious clock, and views how regularly and constantly the *Index* moves upon the dial-plate, how orderly and how distinctly the hours are struck, how the alarum-bell rouses mens attention at determinate times, and the chimes do at other times delight his ear by their harmonious strokes; though this man, I say, cannot certainly decide a controversy, that may arise from clock-makers; whereof one may affirm, that the machine is moved barely by *weights;* another may derive its motions from those of a *pendulum;* and a third may dissent from both those, and substitute to weights a *spring:* . . . this being not founded upon the truth of this or that *Hypothesis,* about the latent structure of the engine, but upon the inspection

of the curious and regular Phaenomena themselves, that it manifestly exhibits. (Boyle, *The Christian Virtuoso,* in *Works,* 6:721–22)

148. Boyle, *Works,* 2:21–22.

149. There may be things, which though we might understand well enough, if God . . . did make it his work to inform us of them, yet we should never of ourselves find out those truths. As an ordinary watch-maker may be able to understand the curiousest contrivance of the skilfullest artificer, if this man take care to explain his engine to him, but would never have understood it, if he had not been taught. (Boyle, *Works,* 2:46)

150. "It is infinitely more credible that we, miserable beings, wanderers from the path of bright knowledge, should be incapable of comprehending the method of the Supreme Artificer in his wondrous and wise machinery, than that a coarse smith should be but a rude admirer of the exquisitely elegant workmanship of a watch." (Thomas Sydenham, *The Works,* ed. R. G. Latham, 2 vols. [London, 1848–50], 2:84); and "Whilst we are destitute of senses acute enough to discover the minute particles of bodies, and to give us ideas of their mechanical affections, we must be content to be ignorant of their properties and ways of operation" (John Locke, *Essay concerning Human Understanding,* 2:217).

151. Ibid., 2:58.

152. Our faculties carry us no further towards knowledge and distinction of substances, than a collection of *those sensible ideas which we observe in them;* which, however made with the greatest diligence and exactness we are capable of, yet is more remote from the true internal constitution from which those qualities flow, than, as I said, a countryman's idea is from the inward contrivance of that famous clock at Strasburg, whereof he only sees the outward figure and motions. (Ibid., 2:64)

153. Ibid., 1:403–4.

154. Peter Shaw, "A General Preface," in *The Philosophical Works of the Honourable Robert Boyle,* 3 vols. (London, 1725), 1:1.

155. See, for example, Boyle, *An Hydrostatical Discourse Occasioned by the Objections of the Learned Dr. Henry More,* in *Works,* 3:597.

156. See, for example, Boyle, *About the Excellency and Grounds of the Mechanical Hypothesis,* in *Works,* 4:67–78; or *The Christian Virtuoso,* in *Works,* 5:513–14.

157. Boas, "The Establishment of the Mechanical Philosophy," p. 485.

158. Boyle, *The Christian Virtuoso,* in *Works,* 5:513 (quoted on p. 55).

159. Boyle, *About the Excellency and Grounds of the Mechanical Hypothesis,* in *Works,* 4:70.

160. Ibid., 4:74.

161. Boyle, *The Excellency of Theology Compared with Natural Philosophy* in *Works,* 4:49.

162. A few examples, chosen at random: "According to our doctrine, the world we live in is not a moveless or indigested mass of matter, but an 'Αυτόματον, or *self-moving engine,* wherein the greatest part of the common

matter of all bodies is always . . . in motion" (Boyle, *An Excursion about the Relative Nature of Physical Qualities* [1666], in *Works,* 3:34); "Nor must we look upon the universe that surrounds us as upon a moveless and undistinguished heap of matter, but as upon a great engine" (ibid., 3:24); and ". . . that great engine we call the world" (Boyle, *An Hydrostatical Discourse,* in *Works,* 3:596).

> 163. To say, that though in natural bodies, whose bulk is manifest and their structure visible, the mechanical principles may be usefully admitted, that are not to be extended to such portions of matter, whose parts and texture are invisible; may perhaps look to some, as if a man should allow, that the laws of mechanism may take place in a town-clock, but cannot in a pocket-watch. (Boyle, *About the Excellency and Grounds of the Mechanical Hypothesis,* in *Works,* 4:71)

164. Boyle, *An Excursion about the Relative Nature of Physical Qualities,* in *Works,* 3:24.

165. For example, see Boyle, *Works,* 2:175, 180; 3:685; and 5:84.

> 166. A silent and a striking watch are but one species to those who have but one name for them: but he that has the name *watch* for one, and *clock* for the other, and distinct complex ideas to which those names belong, to *him* they are different species. It will be said perhaps, that the inward contrivance and constitution is different between these two, which the watchmaker has a clear idea of. And yet it is plain they are but one species to him, when he has but one name for them. For what is sufficient in the inward contrivance to make a new species? There are some watches that are made with four wheels, others with five; is this a specific difference to the workman? Some have strings and physies [fusees], and others none; some have the balance loose, and others regulated by a spiral spring, and others by hogs' bristles. (John Locke, *Essay concerning Human Understanding.* 2:88)

167. Ibid., p. 88.

168. Robert Hooke, *The Posthumous Works,* ed. Richard Waller (London, 1705), p. 39.

169. Power, *Experimental Philosophy,* pp. 5, 6, and 193.

170. Glanvill, *Scepsis scientifica,* p. 92.

171. Hale, *The Primitive Origination of Mankind,* p. 48.

172. Thomas Burnet, *The Sacred Theory of the Earth* (1684) (Carbondale, Ill., 1965), pp. 156 and 209.

173. Boyle, *Works,* 2:23, 175, and 180; 3:685; 5:84, 136, 230, and 442; and 1:22.

174. Webster, "Henry More and Descartes," p. 360.

175. Henry More, Letter to Descartes, Cambridge, 11 December 1648, quoted from Cohen, "Descartes and Henry More on the Beast Machine," p. 50.

176. A statement by Ralph Cudworth, published thirty years later, is representative:

> Though in truth all those, who deny the *Substantiality* of *Sensitive Souls,* and will have *Brutes* to have nothing but *Matter* in them, ought consequently

according to Reason, to do as *Cartesius* did, deprive them of all *Sense*. But on the contrary, if it be evident from the Phaenomena, that Brutes are not meer *senseless Machins* or Automata, and only like *Clocks* or *Watches,* then ought not popular *Opinion* and *Vulgar Prejudice* so far to prevail with us, as to hinder Our Assent, to that which sound Reason and Philosophy clearly dictates, that therefore they must have something more than Matter in them. (Ralph Cudworth, *The True Intellectual System of the Universe* [1678] [Stuttgart, 1964], p. 863)

177. Cohen, "Descartes and Henry More on the Beast Machine," pp. 49–50. The same view is expressed in Shugg, "The Cartesian Beast Machine in English Literature," p. 292.

178. Boyle, *Of the High Veneration Man's Intellect Owes to God* (1685), in *Works,* 5:136.

179. Boyle, *A Free Inquiry into the Vulgarly Received Notion of Nature* in *Works,* 5:236, and *About the Excellency . . . of the Mechanical Hypothesis,* in *Works,* 4:71.

180. Boyle then adds:

Though the body of a man be indeed an engine, yet there is united to it an intelligent being (the rational soul or mind) which is capable . . . to discern . . . what may conduce to the welfare of it, and . . . to do many of those things it judges most conducive to the welfare and safety of the body. . . . So that man is not like a watch, or an empty boat, where there is nothing but what is purely mechanical; but like a manned boat, where, besides the mechanical part . . . there is an intelligent being, that takes care of it, and both steers it, or otherwise guides it, and when need requires, trims it. (Ibid., 5:236)

181. He, that knows the structure and other mechanical affections of a watch, will be able by them to explicate the phaenomena of it, without supposing, that it has a soul or life to be the internal principle of its motions or operations; so he that does not understand the mechanism of a watch, will never be enabled to give a rational account of the operation of it, by supposing . . . that the watch is an . . . animal, or living body, and endowed with a soul. (Ibid., 5:245) (italics added)

182. Neither the mechanism of a human body, nor that of very considerable parts of it, is to be judged of only by the structure of the visible parts . . . or even by the texture of those fluid ones, which are to be found in the vessels and cavities of a dead body, when dissected. . . . For I take the body of a living man to be a very compounded engine, such as mechanicians would call *hydraulico-pneumatical;* many of whose functions, (if not the chiefest) are performed, not by the blood and other visible fluids barely as they are liquors, but partly by their circulating and other motions, and partly by a very agile and invisible sort of fluids, called spirits, vital and animal, and partly perhaps, (as I have sometimes guessed) by little springy particles; and perhaps too, by somewhat, that may be called the vital portion of the air; and by things

> analogous to local ferments; the important operations of all which are wont to cease with life, and the agents themselves are not to be discerned in a dead body. (Boyle, *A Disquisition about the Final Causes of Natural Things* [1688], in *Works,* 5:442)

183. See, for example, Rosenfield, "Un chapitre de l'histoire de l'animal machine"; and Shugg, "The Cartesian Beast Machine in English Literature."

184. The following quotations show how representative authors formulated the argument, and the repetitiveness of the selections underscores the extent to which these authors agreed about the substance and importance of the argument.

Robert Boyle in the early 1650s:

> When . . . I see in a curious clock, how orderly every wheel and other parts perform its own motions . . . , I do not imagine, that any of the wheels, etc. or the engine itself is endowed with reason, but commend that of the workman, who framed it so artificially. So when I contemplate the actions of those several creatures, that make up the world, I do not conclude . . . the vast engine itself to act with reason or design, but admire and praise the most wise Author. (Boyle, *The Usefulness of Natural Philosophy,* 2 vols., London, 1772 in *Works,* 2:40)

Joseph Glanvill in 1665:

> To suppose a *Watch,* or any other the most curious *Automaton* by the blind hits of *Chance,* to perform diversity of orderly *motions,* to shew the *hour, day* of the *Month, Tides, age* of the *Moon,* and the like, with an unparallel'd exactness, and all without the regulation of Art; this were the more pardonable absurdity [compared to assuming that our bodies owe their existence not to divine creator but to chance]. (Glanvill, *Scepsis scientifica,* p. 32)

Matthew Hale in 1677:

> If I should see a curious Watch, curiously wrought, graved, and enameled, and should observe the exact disposition of the Spring, the String, the Wheels, the Ballance, the Index, and by an excellent, orderly, regular Motion described, discovering the Hour of the Day, Day of the Month, and divers other regular and curious Motions: Or if I should see such a goodly *Machina* as some ascribe to *Archimedes* . . . , I should most reasonably conclude, that these were neither Casual nor simply Natural Productions, but they were the Work of some intelligent curious Artist, that by design, intention and appropriation wrought and put in order and motion these curious *Automata.* (Hale, *The Primitive Origination of Mankind,* pp. 324–25)

Boyle in 1690:

> A Machine so immense, so beautiful, so well contrived, and, in a word, so admirable as the world, cannot have been the effect of mere chance, or the tumultuous justlings and fortuitous concourse of atoms, but must have been

produced by a cause exceedingly powerful, wise, and beneficient. (Boyle, *The Christian Virtuoso,* in *Works,* 5:519)

And John Ray in 1691:

For . . . as it argues and manifests more skill by far in an artificer to be able to frame both clocks and watches, and pumps and mills, and granadoes and rockets, than he could display in but making one of those sorts of engines, so the Almighty discovers more of His wisdom in forming such a vast multitude of different sorts of creatures, and all with admirable and irreprovable art, than if He had created but a few; for this declares the greatness and unbounded capacity of His understanding. (John Ray, *The Wisdom of God Manifested in the Works of the Creation* (1691), quoted from Westfall, *Science and Religion in Seventeenth-Century England,* pp. 45–46)

185. According to Sir Kenelm Digby in 1645:

He were an improvident clockemaker, that should have cast his worke so, as when it were wound up and going, it would require the masters hand att every houre to make the hammer strike upon the bell. Lett us not then too familiarly, and irreverently ingage the Almighty Architect his immediate handyworke in every particular effect of nature. (Sir Kenelm Digby, *Two Treatises* [Paris, 1645], p. 227)

And in 1651 Henry More said:

There is an Impulsive cause and End of their working, though unknown to them, yet not unknown to the Authour of them. As in the orderly motion of a Watch, the Spring knows not the end of its Motion, but the Artificer doth. Yet the watch moves, and orderly too, and to a good End. (Henry More, *The Second Lash of Alazonomestix* [London, 1651], quoted from Robert A. Greene, "Henry More and Robert Boyle on the Spirit of Nature," *Journal of the History of Ideas* 23 [1962]: 458)

186. Henry Power offered this argument:

For would it not even in a common Watch maker (that has made a curious Watch for some Gentlemen or other, to shew him the rarity of his Art) be great indiscretion, and a most impudent act, and argue also a dislike of his own work, to pluck the said Watch in pieces before every wheel therein had made one revolution at least? Now the *Apogaeum* (if it move equally, as it hath hitherto done) will not perfect one Revolution under 20,000 years, whereof there is but one Quadrant yet spent, and 15,000 years are yet to come. (Power, *Experimental Philosophy,* pp. 189–90)

187. Methinks we may, without absurdity, conceive, that God, of whom in the scripture it is affirmed, *That all his works are known to him from the beginning,* having resolved, before the creation, to make such a world as this of ours, did divide (at least if he did not create it incoherent) that matter, which he had provided, into an innumerable multitude of very variously figured corpuscles, and both connected those particles into such textures or

particular bodies, and placed them in such situations, and put them into such motions, that by the assistance of his ordinary preserving concourse, the phaenomena, which he intended should appear in the universe, must as orderly follow, and be exhibited by the bodies necessarily acting according to those impressions or laws, though they understand them not at all, as if each of those creatures had a design of self-preservation, and were furnished with knowledge and industry to prosecute it; and as if there were diffused through the universe an intelligent being, watchful over the publick good of it, and careful to administer all things wisely for the good of the particular parts of it, but so far forth as is consistent with the good of the whole, and the preservation of the primitive and catholick laws established by the supreme cause; as in the formerly mentioned clock of *Strasburg,* the several pieces making up that curious engine are so framed and adapted, and are put into such a motion, that though the numerous wheels, and other parts of it, move several ways, and that without any thing either of knowledge or design; yet each performs its part in order to the various end, for which it was contrived, as regularly and uniformly as if it knew and were concerned to do its duty. And the various motions of the wheels and other parts concur to exhibit the phaenomena designed by the artificer in the engine, as exactly as if they were animated by a common principle, which makes them knowingly conspire to do so, and might, to a rude Indian, seem to be more intelligent than *Conradus Diasypodius* himself, that published a description of it; wherein he tells the world, that he contrived it, who could not tell the hours, and measure time so accurately as his clock. (Boyle, *The Usefulness of Natural Philosophy,* in *Works,* 2:39)

188. Boyle, *The Excellency of Theology Compared with Natural Philosophy,* in *Works,* 4:49.

189. It much more tends to the illustration of God's wisdom, to have so framed things at first, that there can seldom or never need any extraordinary interposition of his power. And, as it more recommends the skill of an engineer to contrive an elaborate engine so, as that there should need nothing to reach his ends in it but the contrivance of parts devoid of understanding, than if it were necessary, that ever and anon a discreet servant should be employed to concur notably to the operations of this or that part, or to hinder the engine from being out of order; so it more sets off the wisdom of God in the fabric of the universe, that he can make so vast a machine perform all those many things, which he designed it should, by the meer contrivance of brute matter managed by certain laws of local motion and upheld by his ordinary and general concourse, than if he employed from time to time an intelligent overseer, such as nature is fancied to be, to regulate, assist, and controul the motions of the parts: in confirmation of which you may remember, that the later poets justly reprehended their predecessors for want of skill in laying the plots of their plays, because they often suffered things to be reduced to that pass, that they were fain to bring some deity (θεὸς ἀπό μηχανῆς) upon the stage, to help them out. (Boyle, *A Free Inquiry into the Vulgarly Received Notion of Nature,* in *Works,* 5:162–63)

190. They seem to imagine the world to be after the nature of a puppet, whose contrivance indeed may be very artificial, but yet is such, that almost every particular motion the artificer is fain (by drawing sometimes one wire or string, sometimes another) to guide and oftentimes over-rule the actions of the engine; whereas, according to us, it is like a rare clock, such as may be that at *Strasburgh,* where all things are so skilfully contrived, that the engine being once set a moving, all things proceed, according to the artificer's first design, and the motions of the little statues, that at such hours performs these or those things, do not require, like those of puppets, the peculiar interposing of the artificer, or any intelligent agent employed by him, but perform their functions upon particular occasions, by virtue of the general and primitive contrivance of the whole engine. (Ibid., 5:163)

191. The difficulty we find to conceive, how so great a fabrick, as the world, can be preserved in order, and kept from running again to a chaos, seems to arise from hence; that men do not sufficiently consider the unsearchable wisdom of the divine architect, or Δημιοῦργος (as the scripture styles him) of the world, whose piercing eyes were able to look, at once, quite through the universe, and take into his prospect both the beginning and end of time: so that perfectly fore-knowing what would be the consequences of all the possible conjunctures of circumstances, into which matter, divided and moved according to such laws, could, in an automaton so constituted as the present world is, happen to be put: there can nothing fall out, unless when a miracle is wrought, that shall be able to alter the course of things, or prejudice the constitution of them, any further, than he did from the beginning foresee, and think fit to allow. (Ibid., 5:190)

192. And though I think it probable, that, in the conduct of that far greatest part of the universe which is merely corporeal, the wise Author of it does seldom manifestly procure a recession from the settled course of the universe . . . ; yet, where men, who are creatures, that he is pleased to endow with free wills, . . . are concerned; I think he has, not only sometimes by those signal and manifest interpositions we call miracles, acted by a supernatural way, but, as the sovereign Lord and Governor of the world, doth divers times (and perhaps oftener than mere philosophers imagine) give, . . . divers such determinations to the motions of parts in those [human] bodies, and of others which may be affected by them, as by laws merely mechanical those parts of matter would not have had. (Ibid., 5:215–16)

193. Since man himself is vouchsafed a power to alter, in several cases, the usual course of things, it should not seem incredible, that the latent interposition of men, or perhaps angels, or other causes unthought of by us, should sometimes be employed to the like purposes by God, who is not only the all-wise Maker, but the absolute and yet most just and benign Rector of the universe and of men. (Ibid., 5:216)

194. Henry Stubbe, *The Plus Ultra Reduced to a Nonplus* (London, 1670), p. 172.

195. Cudworth, *The True Intellectual System of the Universe,* preface.
196. Ibid., p. 146.

197. To assert . . . that all the Effects of Nature come to pass by *Material and Mechanical Necessity,* or the mere *Fortuitous Motion of Matter,* [in other words to explain the functioning of the world in terms of a perfect clockwork that, once finished, runs according to its own laws] is a thing no less Irrational than it is Impious and Atheistical. [For] those theists, who Philosophize after this manner, . . . make God to be nothing else in the World but an *Idle Spectator* of the Various Results of the *Fortuitous* and *Necessary Motions* of Bodies; and render his Wisdom altogether Useless and Insignificant, as being a thing wholly Inclosed and shut up within his own breast. (Ibid., p. 148)

200. First, that a machine so immense, so beautiful, so well contrived, and, in a word, so admirable as the world, cannot have been the effect of mere chance, or the tumultuous justlings and fortuitous concourse of atoms, but must have been produced by a cause exceedingly powerful, wise, and beneficient.

Secondly, that this most potent Author, and (if I may so speak) Opificer of the world, hath not abandoned a master-piece so worthy of Him, but does still maintain and preserve it, so regulating the stupendiously swift motions of the great globes, and other vast masses of the mundane matter, that they do not, by any notable irregularity, disorder the grand system of the universe, and reduce it to a kind of chaos, or confused state of shuffled and depraved things.

Thirdly, that as it is not above the ability of the divine author of things, though a single being, to preserve and govern all his visible works, how great and numerous soever; so he thinks it not below his dignity and majesty, to extend his care and beneficence to particular bodies, and even to the meanest creatures; providing not only for the nourishment, but for the propagation of spiders and ants themselves. And indeed, since the truth of this assertion, that God governs the world he has made, would appear (if it did not by other proofs) by the constancy and regularity, and astonishing rapid motions of the vast celestial bodies, and by the long trains of as admirable as necessary artifices, that are employed to the propagation of various sorts of animals, whether viviparous, or oviparous; I see not, why it should be denied, that God's providence may reach to his particular works here below, especially to the noblest of them, man. (Boyle, *The Christian Virtuoso,* in *Works,* 5:519)

201. George Cheyne, *Philosophical Principles of Natural Religion* (1705) (London, 1715), p. 143.
202. Ibid., p. 138.
203. William Derham, *The Artificial Clockmaker: A Treatise of Watch and Clock-work* (London, 1696).
204. William Derham, *Astro-Theology* (1714) (London, 1715), p. 68.

205. By this doctrine, though an artist hath made the spring and wheels, and every movement of a watch, and adjusted them in such a manner as he knew

> would produce the motions he designed, yet he must think all this done to no purpose, and that it is an Intelligence which directs the index and points to the hour of the day. If so, why may not the Intelligence do it, without his being at the pains of making the movements and putting them together? Why does not an empty case serve as well as another? (George Berkeley, *A Treatise concerning the Principles of Human Knowledge* [Dublin, 1710], p. 283)

206. Ibid., p. 285.

207. Isaac Newton, *Mathematical Principles of Natural Philosophy* (1687), ed. W. David, trans. Andrew Motte, 3 vols. (London, 1803), 2:310. The "General Scholium" was first published in the second edition of 1713.

208. Isaac Newton, *The Correspondence,* ed. H. W. Turnbull (Cambridge, 1959–), 3:234.

209. Isaac Newton, *Opticks,* 4th ed. (London, 1730), p. 378, 31st query.

210. Newton, *Mathematical Principles,* pp. 311, 313.

211. Newton, *Opticks,* p. 379.

212. This theme is discussed at length in David Kubrin, "Newton and the Cyclical Cosmos: Providence and the Mechanical Philosophy," *Journal of the History of Ideas* 28 (1967): 325–46.

213. Newton, *Opticks,* pp. 378, 373.

214. David Gregory, Memorandum on conversation with Newton, Cambridge, 5, 6, 7 May 1694, quoted from I. Bernard Cohen, *Introduction to Newton's Principia* (Cambridge, 1971), p. 192.

215. Ibid., 16(?) May 1694.

216. Typical examples are quoted in David Kubrin, "Newton and the Cyclical Cosmos," *Journal of the History of Ideas* 28 (1967): 330, 334, 338.

217. This point has been argued energetically for some time by Stephen G. Brush in, for example, Gerald Holton, *Concepts and Theories in Physical Science,* 2d ed. by Stephen G. Brush (Reading, Mass., 1973), pp. 283–84; and in Stephen G. Brush, "The Development of the Kinetic Theory of Gases," *Archives for the History of the Exact Sciences* 12 (1974): 3–5.

218. Used here is H. G. Alexander, ed., *The Leibniz-Clarke Correspondence* (Manchester, 1956).

219. On Samuel Clarke, see John Gay, "Matter and Freedom in the Thought of Samuel Clarke," *Journal of the History of Ideas* 24 (1963): 85–105; Joel M. Rodney, "Clarke, Samuel," *Dictionary of Scientific Biography* (New York, 1971), 3:294–97.

220. Samuel Clarke, *A Third and Fourth Defense of an Argument to prove the Immateriality and Natural Immortality of the Soul* (1708), 2d ed. (London, 1712), p. 59, quoted from Gay, "Matter and Freedom," pp. 88–89.

221. For entry into the vast literature on this subject, see Steven Shapin, "Of Gods and Kings: Natural Philosophy and Politics in the Leibniz-Clarke Disputes," *Isis* 72 (1981): 187–215.

222. On Newton's role in the debate, see A. R. Hall and M. B. Hall, "Clarke and Newton," *Isis* 52 (1961): 583–85; and A. Koyré and I. B. Cohen, "Newton and the Leibniz-Clarke Correspondence," *Archives internationales d'Histoire des Sciences* 15 (1962): 63–126.

223. Leibniz, Letter to Caroline, Princess of Wales, November 1715, in Alexander, *The Leibniz-Clarke Correspondence,* pp. 11–12.

224. Ibid., Samuel Clarke, First reply to Leibniz, 26 November 1715, pp. 13–14.

225. Ibid., Samuel Clarke, Third reply, p. 34.

Chapter 4. The Clockwork State

1. These references and others are cited in David G. Hale, "Analogy of the Body Politic," in *Dictionary of the History of Ideas,* ed. Philip P. Wiener, 5 vols. (New York, 1973–74), 1:68.

2. Examples are cited in Alexander Demandt, *Metaphern für Geschichte,* (Munich, 1978), see index. Also useful are Ahlrich Meyer, "Mechanische und organische Metaphorik politischer Philosophie," *Archiv für Begriffsgeschichte* 13 (1969): 128–99; and Angus Fletcher, *Allegory: The Theory of the Symbolic Mode* (Ithaca, N.Y., 1964), passim.

3. Detlev Langebek, *Der ander Teil des Regentenbuchs de institia, iure et aequitate* (Wittenberg, 1572), p. 35, cited in Hans L. Stoltenberg, *Geschichte der deutschen Gruppwissenschaft (Soziologie) mit besonderer Beachtung ihres Wortschatzes,* part 1 (Leipzig, 1937), p. 49.

4. See chap. 2, n. 50, above.

5. Henri Duc de Rohan, *Trutina statuum Europae sive principum Christiani orbis interesse* (1634) (Leyden, 1644), pp. 10–11.

6. See Daniel Sennert, *Thirteen Books of Natural Philosophy* (London, 1659), p. 474; see also p. 475.

7. Florentius Schoonhovius, *Emblemata partim Moralia partim etiam Civilia* (Gouda, 1618), no. 7, quoted in A. Henkel and A. Schöne, eds., *Emblemata: Handbuch zur Sinnbildkunst des XVI. und XVII. Jahrhunderts* (Stuttgart, 1967), p. 1341.

8. Diego de Saavedra Fajardo, *Idea de un Principe Politico Christiano* (Munich, 1640), no. 57, quoted in Henkel and Schöne, *Emblemata,* pp. 1341–42.

9. Everything is best understood by its constitutive causes. For as in a watch, or some small engine, the matter, figure, and motion of the wheels cannot well be known, except it be taken insunder and viewed in parts; so to make a more curious search into the rights of states and duties of subjects, it is necessary, I say, not to take them insunder, but yet that they be so considered as if they were dissolved; that is, that we rightly understand what the quality of human nature is, in what matter it is, in what not, fit to make up a civil government, and how men must be agreed amongst themselves that intend to grow up into a well-grounded state. (Thomas Hobbes, *De Cive,* preface, in Hobbes, *The English Works,* ed. Sir William Molesworth, 11 vols. [London, 1839–45], 2:xiv)

10. Nature, the art whereby God hath made and governs the world, is by the *art* of man, as in many other things, so in this also imitated, that it can make an artificial animal. For seeing life is but a motion of limbs, the beginning where-

of is in some principal part within; why may we not say, that all *automata* (engines that move themselves by springs and wheels as doth a watch) have an artificial life? For what is the *heart,* but a *spring;* and the *nerves,* but so many *strings;* and the joints, but so many wheels, giving motion to the whole body, such as was intended by the artificer? *Art* goes yet further, imitating that rational and most excellent work of nature, *man.* For by art is created that great LEVIATHAN called a COMMONWEALTH, or STATE, in Latin CIVITAS, which is but an artificial man. (Thomas Hobbes, *Leviathan,* in Hobbes, *English Works,* 3:ix)

11. The *sovereignty* is an artificial *soul,* as giving life and motion to the whole body; the *magistrates,* and other *officers* of judicature and execution, artificial *joints; rewards* and *punishment,* by which fastened to the seat of the sovereignty every joint and member is moved to perform his duty, are the *nerves,* that do the same in the body natural; the *wealth* and *riches* of all the particular members, are the *strength; salus populi,* the *people's safety, its business; counsellors,* by whom all things needful for it to know are suggested unto it, are the *memory; equity,* and *laws,* and artificial *reason* and *will; concord, health; sedition, sickness;* and *civil war, death.* Lastly, the *pacts* and *covenants,* by which the parts of this body politic were at first made, set together, and united, resemble that *fiat,* or the *let us make man,* pronounced by God in the creation. (Ibid., 3:ix–x)

12. William Penn, *Charter of Liberties and Frame of Government of Pennsylvania* (1682), preface, in W. Keith Kavenagh, ed., *Foundations of Colonial America: A Documentary History,* 3 vols. (New York, 1973), 2:1135.

13. George Savile, First Marquess of Halifax, *The Character of a Trimmer* (1688), in *The Complete Works,* ed. Walter Raleigh (Oxford, 1912), pp. 60, 47, 70.

14. John Locke, *Some Considerations of the Lowering and Raising the Value of Money* (1691), in *The Works,* ed. Edmund Law, 4 vols. (London, 1824), 4:21.

15. Walter Moyle, *Democracy Vindicated* (ca. 1699), in Caroline Robbins, ed., *Two English Republican Tracts* (Cambridge, 1969), p. 225.

16. Adam Smith, *The Theory of Moral Sentiments* (London and New Rochelle, N.Y., 1969), pp. 265, 267.

17. Adam Smith, *The Wealth of Nations,* ed. E. Cannan (New York, 1937), pp. 418, 595, and also, for example, p. 607.

18. "Das Gewichte / welches die Uhr des gemeinen Wesens triebe / sey die Krafft eines lebhaften Geistes" (Daniel Casper von Lohenstein, *Grossmüthiger Feldherr Arminius,* ed. E. M. Szarota, 2 vols. [Bern, 1973], 1:229a); "August alleine war die Unruh in der Uhr des gemeinen Wesens [of imperial Rome]" (Ibid., 2:962b). Other more casual clockwork metaphors are found in Ibid., 1:139 and 296 and 2:399. "Denn in der Uhr der Herrschaft sind die Räthe wol die Räder / der Fürst aber muß nichts minder der Weiser / als das Gewichte; Er die Sonne / jene der Monden seyn" (Daniel Casper von Lohenstein, *George Wilhelms . . . Lob-Schrifft* [Breslau and Leipzig, 1679], p. F2, verso).

19. Wilhelm von Schröder, *Fürstliche Schatz- und Rentkammer* (1686) (Leipzig, 1713), preface, §8.

20. On connections between absolutism and mechanism, see James E. King, *Science and Rationalism in the Government of Louis XIV, 1661–1683* (Baltimore, 1949), esp. pp. 21f., 49–51; Meyer, "Mechanische und organische Metaphorik," pp. 128–99; and Marc Raeff, "The Well-Ordered Police State and the Development of Modernity in Seventeenth- and Eighteenth-Century Europe: An Attempt at a Comparative Approach," *American Historical Review* 80 (1975): 1221–43.

21. Ronald S. Calinger, "Frederick the Great and the Berlin Academy of Sciences, 1740–1766," *Annals of Science* 24 (1968): 239–51.

22. Frederick II of Prussia, *L'Anti-Machiavel,* in: *Studies on Voltaire and the 18th Century,* ed. C. Fleischauer (Geneva, 1958), 5:201; Idem, Letter to Voltaire, 26 December 1737, in *Posthumous Works,* 13 vols. (London, 1789), 6: 229–30.

23. Idem, *L'Anti-Machiavell,* p. 323.

24. Idem, Letter to Voltaire, 24 October 1765.

25. A plausible preceptor in this habit might have been Christian Wolff, who has been described as "in the 18th century the dominant theoretician of absolutism in Germany" (Jürgen von Krüdener, *Die Rolle des Hofes im Absolutismus* [Stuttgart, 1973], p. 20). There is doubt, however, that Frederick had read many of Wolff's works (Werner Frauendienst, *Christian Wolff als Staatsdenker* [Berlin, 1927], p. 40).

26. Frederick II of Prussia, *Considérations sur l'état présent du corps politique de l'Europe* (1736), in *Posthumous Works,* 4:342.

27. Idem, Letter to Voltaire, 8 January 1739.

28. Idem, *Dissertation sur les raisons d'établir ou d'abroger les lois* (1750), in Frédéric le Grand, *Oeuvres,* ed. J. D. E. Preuss, 3 vols. (Berlin, 1846–57), 9:24.

> 29. As all the wheels of a watch correspond to effect the same purpose, which is that of measuring time, so ought the springs of government to be regulated, that all the different branches of administration may equally concur to the greatest good of the state.
>
> The Prince is to the nation which he governs what the head is to the man; it is his duty to see, think, and act for the whole community, that he may procure it every advantage of which he is capable. (Frederick II of Prussia, *Essai sur les formes de gouvernement* [1781], in *Posthumous Works,* 5:13, 15)

30. Johann Wolfgang Goethe, Letter to Charlotte von Stein, Berlin, 17 May 1778, in *Werke* (Weimar, 1888), sec. 4, 3:225.

31. Ernst Brandes, *Betrachtungen über den Zeitgeist in Deutschland in den letzten Dezennien des vorigen Jahrhunderts* (Hannover, 1808), pp. 38 and 50, 55, and 57–58.

32. Etienne Bonnot de Condillac, *Traité des systèmes* (1749), in *Oeuvres philosophiques,* ed. Georges Le Roy, 3 vols. (Paris, 1947–51), 1:208.

33. Jean Jacques Rousseau, *Du Contrat social, ou Essai sur la forme de la république* (Manuscrit de Genève, 1754–55?), in *Oeuvres complètes,* l'Intégrale ed., 3 vols. (Paris, 1971), 2:392, translation from Roger D. Masters, *The Political Philosophy of Rousseau* (Princeton, N.J., 1968), p. 260.

34. Idem, *Contrat social* (1762), in *Oeuvres complètes,* l'Intégrale ed., 2:531.

35. En effet, jamais le gouvernement ne change de forme que quand son ressort usé le laisse trop affaibli pour pouvoir conserver la sienne. Or, s'il se relâchait encore en s'étendant, sa force deviendrait tout à fait nulle et il subsisterait encore moins. Il faut donc remonter et serrer le ressort à mesure qu'il cède: autrement l'Etat qu'il soutient tomberait en ruine. (Rousseau, *Contrat social,* 2:554)

36. Rousseau, *Lettres écrites de la montagne* (1764), *Oeuvres complètes,* l'Intregrale ed. in 3:490.

37. See Rousseau, *l'État de guerre,* in *Political Writings,* ed. F. Watkins (Edinburgh, 1953), 1:298–99, and also *Contrat social,* 1:462.

38. Idem, *Contrat social,* 2:524.

39. J. W. Chapman has shown that modern commentators on Rousseau are about evenly divided on the question of whether he should be viewed as a liberal or an authoritarian; see John W. Chapman, *Rousseau—Totalitarian or Liberal?* (New York, 1968), pp. 74–75.

40. Rousseau, "Economie politique," Diderot, and d'Alembert, *l'Encyclopédie,* in *Political Writings,* ed. Watkins, 1:241–42.

41. Ainsi la volonté du peuple, et la volonté du prince, et la force publique de l'État, et la force particulière du gouvernement, tout répond au même mobile, tous les ressorts de la machine sont dans la même main, tout marche au même but; il n'y a point de mouvements opposés qui s'entre-détruisent, et l'on ne peut imaginer aucune sorte de constitution dans laquelle un moindre effort produise une action plus considérable. Archiméde, assis tranquillement sur le rivage et tirant sans peine à flot un grand vaisseau, me représente un monarque habile, gouvernant de son cabinet ses vastes États, et faisant tout mouvoir en paraissant immobile. (Rousseau, *Contrat social,* 2:546)

42. Friedrich Karl von Moser, *Der Herr und der Diener, geschildert mit patriotischer Freyheit* (Frankfurt, 1759), p. 239.

43. Ludwig Benjamin Martin Schmidt, *Lehre von der Staatswirtschaft,* 2 vols. (Mannheim, 1760).

44. For example, Johann Heinrich Gottlieb von Justi, *Der Grundriss einer guten Regierung* (Frankfurt, 1759), pp. 10–12, 185, 186, 187, 317, and 409–410; and *Gesammelte Politische und Finanzschriften,* 3 vols. (Copenhagen and Leipzig, 1761–64), 2:16.

45. For example, "Ein Staat ist ein einfacher moralischer Körper, dessen Theile den allergenauesten Zusammenhang miteinander haben. Er ist eine Maschine, dessen Räder und Triebfedern sehr wohl in einander passen müssen, wenn die Maschine alle Kräfte und Thätigkeit zeigen soll, deren sie fähig ist" (Justi, *Grundriss einer guten Regierung,* p. 320); or "Der Nahrungsstand ist das Triebwerk in der großen Maschinerie des Staates. Die Gewerbe sind die Räder und Federn; und ein jedes Gewerbe muß darinnen seine behörige Stelle einnehmen, und so viel zu der Bewegung der Maschine beytragen, als zu dem Aufnehmen des Nahrungsstandes und der Wohlfahrt des Staates erforderlich ist" (Justi, *Die Grundfeste zu der Macht und Glückseligkeit der Staaten* [Königsberg and Leipzig, 1760], 1:557).

46. Ein Staat, der mit Unordnung regieret wird, ist allemal, ungeachtet der innerlichen Kräfte, die er hat, sehr schwach. Er kann nicht die Hälfte von seiner Stärke und Thätigkeit zeigen, deren er fähig ist. Allein ein Staat, der mit Ordnung beherrschet wird, ist eine Maschine, die mit allen ihren Kräften spielt, und der keine äußerliche Kraft widerstehen kann, so groß sie auch ist. . . . Ein Staat, welcher die aller vollkommenste Ordnung hätte, würde ganz unüberwindlich seyn, wenn auch seine Kräfte nur mittelmäßig wären. (Justi, *Grundriss einer guten Regierung,* pp. 321–22)

47. "Sodann muß seine hauptsächlichste Vorsorge seyn, unaufhörlich zu wachen, daß diese Ordnung aufrecht erhalten wird. Das ist seine vornehmste, ja man kann sagen, seine einzige Pflicht. Er ist der Regierer von der Maschine des Staatskörpers. Wenn er nun sein unaufhörliches Augenmerk seyn läßt, daß die Maschine in ihrer Ordnung bleibt und alle Theile in ihrem gerechten Verhältniß und Übereinstimmung erhalten werden; so braucht es gar nichts weiter. Die Maschine wird von selbst gehen und alle Kräfte und Thätigkeiten zeigen, deren sie fähig ist. Die große Wissenschaft eines Regenten ist demnach, die Kenntnis und Einsicht von der Ordnung seines Staats; und alles, was zu dieser Ordnung erfordert wird, . . . müssen ihm eben so genau bekannt seyn, als der Directeur einer großen Maschine aller Triebwerke, Räder und Zusammenfügung der Theile auf das vollkommenste kennen muß, wenn er die Maschine zu regieren und ihr vorzustehen im Stande seyn will. (Justi, *Grundriss einer guten Regierung,* p. 329)

48. Justi, *Staatswirtschaft, oder Systematische Abhandlungen aller ökonomischen und Cameralwissenschaften,* quoted in Albrecht Timm, *Kleine Geschichte der Technologie* (Stuttgart, 1964), p. 42.

49. Johann Friedrich von Pfeiffer, *Grundsätze der Universal-Kameral-Wissenschaft,* 2 vols. (Frankfurt, 1783), 1:25.

50. Ibid., pp. 26–27, 102.

51. Ganz schicklich läßt sich der Staatskörper mit einem physischen vergleichen. Alle vereinigte Theilnehmer der Souverainität sind die Glieder, die Gesetzgebende Macht ist das Haupt, die ausübende das Herz, die Freyheit das Blut, das alle Theilnehmer beseelt; der Vollzieher des gemeinen Willens ist der Regent, der das vom Gesetzgeber erfundene Triebwerk der Maschine aufzieht. (Ibid., p. 29)

52. Ibid., p. 35.

53. Ibid., pp. 182–83.

54. August Ludwig von Schlözer, *Allgemeines Stats Recht und Stats Verfassungslehre* (Göttingen, 1793), pp. 3–4.

55. Hierdurch (Gesellschaftsvertrag) ward der Stat eine Maschine, eine aus unendlich vielen Rädern zusammen gesetzte, eine der künstlichsten Maschinen, die Menschen je erfunden haben: und die Kunst, diese Maschinen aufzuziehen, sie zu richten, sie im Gange zu erhalten, und immer vollkommener zu machen, oder mit einem Worte die Statskunst, stieg in dem Maße höher, als sich der menschliche Verstand in den übrigen Wissenschaften

aufklärte und Progressen machte. (Schlözer, *Neuverändertes Rußland, oder Leben Catharina der Zweyten, Kayserinn von Rußland* [Riga and Leipzig, 1767], preface, quoted in Bernd Warlich, "August Ludwig von Schlözer, 1735–1809, zwischen Reform und Revolution," Phil. diss., University of Erlangen-Nuremberg, 1972, p. 246)

56. Der Stat ist eine Maschine, aber darinn unendlich verschieden von allen andern Maschinen, daß dieselbe nicht für sich fortlaufen kan, sondern immer von Menschen, leidenschaftlichen Wesen, getrieben wird, die nicht Maschinenmäßig gestellt werden können. Daher sind zur besten Stats Verwaltung auch die besten Menschen nötig, sonst kan jene unmöglich bestehen. Und diese Maschinen Directeure heißen Regenten, collective der Souverain. (Schlözer, *Allgemeines Stats Recht,* p. 157)

57. Der Stat ist auch hierinnen eine Maschine (hält nicht ewig): seine Räder nützen sich ab, einige werden ganz unbrauchbar, das Triebwerk stockt, und ruft die bessernde Hand des Künstlers herbey. Bald sind neue und mehrere Räder nöthig; es äußern sich neue Kräfte im Stat, die vorhin gar nicht gewirket hatten; . . . Je größer der Stat ist: desto zusammengesetzter ist die Maschine; desto künstlicher ist sie; desto mehr und öfter braucht sie Verbesserungen. (Schlözer, *Neuverändertes Rußland,* in Warlich, *A. L. von Schlözer, zwischen Reform und Revolution,* p. 246)

58. Claude Adrien Helvétius, *Oeuvres complètes,* ed. Lepetit, 3:262ff., quoted in Meyer, "Mechanische und organische Metaphorik," p. 177.

59. Paul-Henri Thiry d'Holbach, *Système social,* 3 vols. (London, 1773), 2:83.

60. Idem, *Ausgewählte Texte,* ed. M. Naumann (Berlin, 1959), p. 210, quoted in Meyer, "Mechanische und organische Metaphorik," pp. 168–69.

61. Holbach, *Système social,* 1:212.

62. Denis Diderot, *Supplément au voyage de Bougainville* (1772), in *Oeuvres complètes,* ed. J. Assezat and M. Tourneux, 20 vols. (Paris, 1875–77), 2:247.

Chapter 6. Rejection of the Clock Metaphor in the Name of Liberty

1. See chapter 3, the section on Determinism versus Free Will, above.

2. "He [Mr. Jackson] heard Sir Isaac Newton also once pleasantly tell the Doctor [Samuel Clarke] that 'he had broke Leibnitz's Heart with his Reply to him' " (William Whiston, *Historical Memoirs of the Life of Dr. Samuel Clarke* [London, 1730], p. 132, quoted in Frank E. Manuel, *A Portrait of Isaac Newton* [Cambridge, 1968], p. 348).

3. "If the universe bears a greater likeness to animal bodies and to vegetables, than to works of human art, it is more probable that its cause resembles the cause of the former than that of the latter, and its origin ought rather to be ascribed to generation and vegetation than to reason or design" (David Hume, *Dialogues concerning Natural Religion,* ed. Norman Kemp Smith, [New York, 1948], pp. 339–40). Hume's discussion of the design argument is on pp. 141–57.

On the disposal of the design argument, see also Robert H. Hurlbutt, *Hume, Newton and the Design Argument* (Lincoln, Neb., 1965) and S. A. Grave, "Hume's Criticism of the Argument from Design," *Revue internationale de philosophie* 30 (1976): 64–78.

4. William Paley, *Natural Theology; or, Evidences of the Existence and Attributes of the Deity* (London, 1802).

5. James Paxton, *Illustrations of Paley's Natural Theology* (Boston, 1827).

6. Francis Darwin, *Life and Letters of Charles Darwin* (London, 1887), 2:210. See also Joel M. Rodney, "Paley, William," in *Dictionary of Scientific Biography* (New York, 1974), 10:276; Francis C. Haber, *The Age of the World: Moses to Darwin* (Baltimore, 1959), p. 105; and idem, "The Darwinian Revolution in the Concept of Time," *Studium Generale* 24 (1971): 289–307.

7. See chap. 1, p. 15.

8. See chap. 2, pp. 49–52.

9. Samuel Johnson, *Dictionary of the English Language,* 2 vols. (London, 1755), s.v. mechanick.

10. See chap. 3, pp. 55–56.

11. Christopher Hill, *Change and Continuity in Seventeenth-Century England* (London, 1974), pp. 255–60.

12. *Oxford English Dictionary* (Oxford, 1970), vol. 6, s.v. mechanic, mechanical, mechanicism.

13. Colette Avignon, "Evelyn, John," in *Dictionary of Scientific Biography* (New York, 1971), 4:495; and Edward Gibbon, "The Portable Gibbon," in *The Decline and Fall of the Roman Empire,* ed. D. A. Saunders (New York, 1977), p. 65.

14. See the section, Determinism versus Free Will, in chapter 3 of this volume.

15. George Berkeley, Letter to Samuel Johnson, 25 November 1729, in *The Works,* ed. A. A. Luce and T. E. Jessop, 9 vols. (Edinburgh, 1948–57), 280–81.

16. Newton's *Principia* (Cambridge, 1713), trans. Andrew Motte as *Sir Isaac Newton's Mathematical Principles* (London, 1729).

17. Samuel Clarke, *A Third and Fourth Defense of an Argument . . . to prove the Immateriality and Natural Immortality of the Soul* (1708) (London, 1712), p. 59.

18. George Cheyne, *Philosophical Principles of Natural Religion* (1705) (London, 1715), p. 138.

> 19. Gorgias hath gone further, demonstrating man to be a piece of clockwork or machine; and that thought or reason are the same thing as the impulse of one ball against another. Cimon hath made noble use of these discoveries, proving, as clearly as any proposition in mathematics, that conscience is a whim, and morality a prejudice; and that a man is no more accountable for his actions than a clock is for striking. (George Berkeley, *Alciphron, or the Minute Philosopher* (1732), in *Works,* ed. Luce and Jessop, 3:51)

20. Colin Maclaurin, *An Account of Sir Isaac Newton's Philosophical Discoveries* (1748), ed. L. L. Laudan (New York, 1968), p. 83.

21. Thomas Hobbes, "Philosophical Rudiments concerning Government and Society," in *The English Works,* ed. Sir William Molesworth (London, 1841), 2:11. Hobbes, *Leviathan,* ed. C. B. MacPherson (Harmondsworth, 1968), p. 186.

22. Thomas Sprat, *The History of the Royal Society of London* (London, 1667), p. 73.

23. George Savile, First Marquess of Halifax, *A Rough Draught of a New Model at Sea* (1694), in *The Complete Works,* ed. Walter Raleigh (Oxford, 1912), p. 172.

24. Algernon Sidney, *Discourses on Government* (ca. 1680), 3 vols. (New York, 1805), esp. vol. 2.

> 25. If it be desir'd to know the immediat cause of all this free writing and free speaking, there cannot be assign'd a truer than your own mild, and free, and human government; it is the liberty, Lords and Commons, which your own valorous and happy counsels have purchast us, liberty which is the nurse of all great wits; this is that which hath ratify'd and enlightn'd our spirits like the influence of heav'n; this is that which hath enfranchis'd, enlarg'd and lifted up our apprehensions degrees above themselves. Ye cannot make us now lesse capable, lesse knowing, lesse eagerly pursuing of the truth, unlesse ye first make your selves, that made us so, lesse the lovers, lesse the founders of our true liberty. We can grow ignorant again, brutish, formall, and slavish, as ye found us; but you then must first become that which ye cannot be, oppressive, arbitrary, and tyrannous, as they were from whom ye have free'd us. . . . Give me the liberty to know, to utter, and to argue freely according to conscience, above all liberties. (John Milton, *Areopagitica; a Speech . . . For the Liberty of Unlicenc'd Printing, to the Parliament of England* (1644), in *The Works,* ed. Frank A. Patterson [New York, 1931] 4:345)

26. Voltaire, *Philosophical Letters,* trans. Ernest Dilworth (Indianapolis, 1961), p. 31.

27. Ibid., p. 39.

28. "One nation there is also in the world that has for the direct end of its constitution political liberty" (Charles Louis de Montesquieu, *The Spirit of the Laws* [1748], trans. Thomas Nugent [New York, 1949], p. 151).

29. "Dies Ideal einer vorzüglich glücklichen Regierungsform ist bekanntlich mehr als Ideal: England hat sie wirklich, Rom in seiner 1sten Periode hatte sie zum Teil" (August Ludwig Schlözer, *Allgemeines Stats Recht und Stats Verfassungslehre* [Göttingen, 1793], p. 155).

> England liefert uns ein Muster von dergleichen Regierung [mixed government], und ich gestehe, daß wenn menschliche Einrichtungen das Glück der Nationen zu machen geschickt sind, es in dieser zu finden seyn müsse, allwo weise Gesetze allen Gliedern der Gesellschaft befehlen, wo der König selbst ihnen unterworfen ist, und bloß das glückliche Vorrecht hat, Gutes zu thun;

wo die Gesetze gleichsam den allgemeinen Willen der Gesellschaft ausdrücken, wo die Person, die Freyheit, das Eigenthum jedes einzelnen Menschen heilig ist, und durch keine Macht ungestraft benachtheiliget werden kann. (Johann Friedrich von Pfeiffer, *Grundsätze der Universal-Cameral-Wissenschaft,* 2 vols. [Frankfurt, 1783], 1:26)

Das ist die gesetzerzeugende Hingebung an das ewige Interesse des Staates, welche diese glückliche Insel durch die furchtbarsten Krisen dieses Jahrhunderts glücklich hindurch geführt hat, deren kleinste schon hinreichen würde, einen Continental-Staat über den Haufen zu werfen. Das erhebt England zum ersten aller Christlichen Staaten; denn die Gegenseitigkeit, die ewige Wechselwirkung zwischen der Freyheit und dem wahren Gesetze, die Hingebung des Einzelnen an das Ganze auf Leben und Tod, ist durch das Christentum, und durch keine andere Fügung der Umstände, in die Welt gekommen. (Adam Heinrich Müller, *Die Elemente der Staatskunst* [Berlin, 1809], ed. Jakob Baxa, 2 vols. [Vienna and Leipzig, 1922], 1:274–75)

30. The attentive reader may ask where the Netherlands, although Continental but with their characteristically liberal and democratic traditions, fit into the scheme. Bearing this question in mind all along, I have not come upon any material to suggest that developments in Holland were different from those in other Continental countries. A substantial answer to the question, however, would require a detailed study of the relationship between literature and technology in the Netherlands from the sixteenth to the eighteenth centuries.

31. Voltaire, *A Philosophical Dictionary,* 6 vols. (London, 1824), 3:256, s.v. Free-Will.

32. For Rousseau's attitude on the principle of freedom, see chap. 4, p. 110 above. On the inconsistent tendency of French philosophes to believe both in the man-machine and in freedom, see Aram Vartanian, "Necessity or Freedom? The Politics of an Eighteenth-Century Metaphysical Debate," *Studies in Eighteenth-Century Culture* 7 (1978): 153–74.

33. Denis Diderot, Letter to Landois, 29 June 1756, in John Robert Loy, *Diderot's Determined Fatalist* (New York, 1950), p. 134; see also Lester Crocker, *Diderot's Chaotic Order: Approach to Synthesis* (Princeton, N.J., 1974), pp. 35–44.

34. Julien Offray de La Mettrie, *L'Homme machine,* ed. Aram Vartanian (Princeton, N.J., 1960), pp. 94–95.

35. Paul-Henry Thiry d'Holbach, *System of Nature, or, Laws of the Moral and Physical World* (1770), trans. H. D. Robinson (Boston, 1853), pp. 101, 103, 40.

36. Christian August Crusius, *Entwurf der notwendigen Vernunft-Wahrheiten, wiefern sie den zufälligen entgegengestellt werden* (1745) (Leipzig, 1753), p. 759.

37. Leonhard Euler, *Letters . . . on different subjects in Natural Philosophy addressed to a German Princess,* ed. David Brewster, 2 vols. (New York, 1838), 1:280.

38. Ibid., pp. 282, 284.

39. Ibid., p. 282.

> 40. Would it not, then, be ridiculous to expect that a watch should point to any other hour than what it actually does, and to think of punishing it on that account? Would it not be absurd to fly into a passion at a puppet because, after several other gestures, it had turned its back to us? . . . The machine itself has no interest in what passes; the artist . . . is alone responsible for the defects of a clumsy and awkward machine; the machine itself is perfectly innocent. (Ibid., p. 283)

41. Ibid., p. 284.
42. Ibid., p. 288.

> 43. Imagine to yourself a musical clock; such a clock once regulated, all the motions which it performs, and the airs which it plays, are produced in virtue of its construction, without any fresh application of the hand of the master, and in that case we say it is done mechanically. If the artist touches it, by changing the notch, or the cylinder, which regulates the airs, or by winding it up, it is an external action, which, not being founded on the organization of the machine, no longer appertains to it. And if God, as Lord of the universe, should change immediately any thing in the course of successive events, this change would no longer appertain to the machine: it would then be a *miracle.* (Ibid., p. 289)

44. Ibid., p. 290.
45. Lewis White Beck, "Introduction to Immanuel Kant," in *Critique of Practical Reason,* ed. L. W. Beck (Chicago, 1949), p. 28.
46. Ibid., pp. 118–19.

> 47. It is a wretched subterfuge to seek an escape in the supposition that the *kind* of determining grounds of his [the criminal's] causality according to natural law agrees with a *comparative* concept of freedom. According to this concept, what is sometimes called "free effect" is that of which the determining natural cause is internal to the acting thing. For example, that which a projectile performs when it is in free motion is called by the name "freedom" because it is not pushed by anything external while in flight. Or, another example: we call the motion of a clock "free movement" because it moves the hands itself, which need not be pushed by an external force. (Ibid., p. 201)

48. Ibid., p. 203.
49. Ibid., p. 206.
50. Immanuel Kant, *Kritik der Urteilskraft* (1790), in *Sämtliche Werke,* ed. Karl Vorländer, 10 vols. (Leipzig, 1920–29), 2:237.
51. Immanuel Kant, *Grundlegung zur Metaphysik der Sitten* (1785), in *Sämtliche Werke,* 3.2:168.
52. Johann Gottlieb Fichte, *Beitrag . . . über die französische Revolution* (1793), in *Gesamtwerke der Bayerischen Akademie der Wissenschaften,* ed. R. Lauth and H. Jakob (Stuttgart, 1964), 1.1:253.
53. Ernst Brandes, *Betrachtungen über den Zeitgeist in Deutschland in den letzten Dezennien des vorigen Jahrhunderts* (Hannover, 1808), pp. 38, 50, 59, and

63. As a condemnation of the mindlessness of the Prussian machine state by an Anglophile Hanoverian government official, this book is unsurpassed.

54. Georg Wilhelm Friedrich Hegel, *Die Verfassung Deutschlands* (1802), in *Sämtliche Werke,* ed. Georg Lasson (Leipzig, 1928–), 7:28, 30.

55. Brandes, *Betrachtungen über den Zeitgeist in Deutschland,* pp. 60–61.

56. Adam Heinrich Müller, *Die Elemente der Staatskunst* (1809), ed. Jakob Baxa, (Vienna and Leipzig, 1922), 1:15–16.

57. Strong statements can be found in the writings of Karl Marx and Friedrich Engels, who called for the forcible destruction of the machine state. See Ahlrich Meyer, "Mechanische und organische Metaphorik politischer Philosophie," *Archiv für Begriffsgeschichte* 13 (1969): 188–92.

58. Alexander Demandt, *Metaphern für Geschichte* (Munich, 1978), pp. 271–76.

59. Clocks also lost their popular appeal. Discussing the complex indications of the famous old towerclocks of Lyon, Nuremberg, Venice, and other cities, an observer of contemporary (1762) technology reported:

> Man sahe ehemals welche [towerclocks], an denen das ganze Kalenderwesen, mit dem Auf-und Untergange, mit den Bewegungen des Mondes und andren dahin einschlagenden Dingen angebracht war. Heutzutage hat sich die Welt davon überzeugt, daß ein einziger Kalender die Stelle aller dieser fürchterlichen und kostbaren Räderwerke vertritt, und nunmehr schränkt sich unsere Uhrmacherkunst immer mehr und mehr auf das Einfache und auf eine vollkommene Genauigkeit ein. (J. S. Halle, *Werkstätte der heutigen Künste* [Brandenburg, 1762], 2:246, in Albert Protz, *Mechanische Musikinstrumente* [Kassel, 1940], p. 28)

Protz also reports that no carillons were built in Germany from 1721 to 1893 and that even King Frederick II of Prussia, with all his mechanistic attitudes, ordered the carillons of the Berlin parochial church to be silenced whenever he was in his city palace (Ibid., p. 34).

60. Good surveys of relevant material are: H. E. Barnes, "Representative Biological Theories of Society," *Sociological Review* 17 (1925): 120–30, 182–94, and 294–300, and 18 (1926): 100–105, 231–43, and 306–14; David C. Hale, "Analogy of the Body Politic," in *Dictionary of the History of Ideas,* ed. Philip P. Wiener (New York, 1968), 1:67–70; Meyer, "Mechanische and organische Metaphorik," pp. 128–99, esp. pp. 154–63 and 193–99; and Demandt, *Metaphern für Geschichte,* pp. 55–123.

61. Kant, *Kritik der Urteilskraft,* in *Sämtliche Werke,* 2.2:212.

62. Fichte, *Beitrag . . . über die französische Revolution,* in *Gesamtwerke,* 1.1:249.

63. Friedrich Schiller, *Über die ästhetische Erziehung des Menschen* (1793–94), in *Sämtliche Werke,* DTV ed., 20 vols. (Munich, 1965–66), 19:18.

64. G. W. F. Hegel, *Differenz des Fichteschen und Schellingschen Systems der Philosophie* (1801), ed. G. Lasson (Hamburg, 1962), p. 69.

65. See Hermann Heller, *Hegel und der nationale Machtstaatsgedanke in Deutschland* (1921) (Aalen, 1963), pp. 17–18.

1. See chap. 6, n. 59.

2. Alexander Demandt, *Metaphern für Geschichte* (Munich, 1978), pp. 301–11, esp. pp. 302–3.

3. Aristotle, *Politics,* trans. H. Rackham (London and Cambridge, Mass., 1959), 1295:b38, quoted from Demandt, *Metaphern,* p. 304.

4. Polybius, *The Histories,* trans. W. R. Paton, 6 vols. (London and Cambridge, Mass., 1954), 6:10, quoted from Demandt, *Metaphern,* p. 304. Demandt also cites similar balance imagery in Plutarch, Livy, Cicero, Ovid, Lucian, Cassianus, and Ausonius.

5. Useful surveys of the early history of the concept of balance of power are Ernst Kaeber, *Die Idee des europäischen Gleichgewichts in der publizistischen Literatur vom 16. bis zur Mitte des 18.Jh.* (Berlin, 1907); Herbert Butterfield, "The Balance of Power," in H. Butterfield and M. Wight, eds., *Diplomatic Investigations* (London, 1966), pp. 132–48; Martin Wight, "The Balance of Power," ibid., pp. 149–75; M. S. Anderson, "Eighteenth-Century Theories of the Balance of Power," in Ragnhild Hatton and M. S. Anderson, eds. *Studies in Diplomatic History: Essays in Memory of David Bayne Horn* (London, 1970), pp. 183–98; Herbert Butterfield, "Balance of Power," in *Dictionary of the History of Ideas,* ed. P. P. Wiener, 5 vols. (New York, 1973), 1:179–88; and Moorhead Wright, ed., *Theory and Practice of the Balance of Power, 1486–1914: Selected European Writings* (London, 1975), esp. the Introduction and Selected Bibliography.

> 6. And lastly [God] hath erected your seate upon a high hill or sanctuarie, and put into your hands the ballance of power and justice, to peaze and counterpeaze [appease and counterpoise] at your will the actions and counsels of all the Christian Kingdomes of your time. (Geffray Fenton, *The History of Guicciardin . . . reduced into English* [1579], epistle dedicatorie to the Queen, pt. 4, in Wight, "Balance of Power," pp. 163–64)

See also Butterfield, "Balance of Power," pp. 135–37.

> 7. It is first to be considered that this part of Christendom is balanced betwixt the three Kings of Spain, France and England; as the other part is betwixt the Russian, the Kings of Poland, Sweden and Denmark. For as for Germany, which if it were entirely subject to one Monarchy, would be terrible to all the rest; so being divided betwixt so many Princes and those of so equal power, it serves only to balance itself, and entertain easy war with the Turk; while the Persian withholds him in a greater. (Sir Thomas Overbury, *Observations in His Travels* [1609], in *Stuart Tracts, 1603–1693,* ed. C. H. Firth [London, 1903], p. 227, quoted in Wight, "Balance of Power," p. 152)

> 8. There sat she as an heroical princess and umpire betwixt the Spaniards, the French, and the Estates; so as she might well have used that saying of her father, *Cui adhereo, praeest,* that is "the party to which I adhere getteth the upper hand." And true it was which one hath written, that France and Spain

> are as it were the scales in the balance of Europe, and England the tongue or holder of the balance. (William Camden, *History of England,* 3d ed. [1675], p. 223, quoted in Wight, "Balance of Power," p. 159)

9. During that triumvirate of kings, King Henry the Eighth of England, Francis the First of France, and Charles the Fifth Emperor, there was such a watch kept, that none of the three could win a palm of ground, but the other two would straight ways balance it, either by confederation, or, if need were, by a war. (Francis Bacon, "On Empire," in *Essays, or Councils Civil and Moral* [before 1625], in *The Works,* ed. J. Spedding, R. L. Ellis, and D. D. Heath, 14 vols. [London, 1857–74], 6:420)

10. Wight, "Balance of Power," p. 164.

11. *France* and *Spain* became the Rivals for the Universal Monarchy, and our third Power, tho in it self less than either of the other, hapned to be Superiour to any of them, by that choice we had of throwing the Scales on that side to which we gave our Friendship. (George Savile, First Marquess of Halifax, *The Character of a Trimmer,* in *The Complete Works,* ed. Walter Raleigh, [Oxford, 1912], p. 87); and,

> The forming of two such powers [France and Austria ca. 1500], in Europe, made it the interest of all other princes and states, to keep as much as possible in balance between them. And here began that principle of English policy to be established, which, however true and wise in itself, has hardly ever been truly and wisely pursued. (Henry Saint-John, First Viscount Bolingbroke, *The Works,* 4 vols. [London, 1844], 1:215)

12. Frederick H. Gareau, ed., *The Balance of Power and Nuclear Deterrence: A Book of Readings* (Boston, 1962), p. 99.

13. J. Dumont, *Corps universel diplomatique du droit des gens* (The Hague, 1726–31), 8, pt. 1, p. 393, quoted in Anderson, "Eighteenth-Century Theories of the Balance of Power," p. 184; see also Wight, "Balance of Power," p. 153.

14. *Two Essays on the Balance of Europe,* (London, 1720), in *A Collection of Scarce and Valuable Tracts* (London, 1809–15), 13:770, quoted in Anderson, "Eighteenth-Century Theories of the Balance of Power," p. 183.

15. Gasparo Contarini, *The Commonwealth and Government of Venice* (1543) (London, 1599), p. 15, quoted in Zera S. Fink, *The Classical Republicans: An Essay in the Recovery of a Pattern of Thought in Seventeenth Century England* (Evanston, Ill.: 1945), p. 34.

16. Details can be found, for example, in Stanley Pargellis, "The Theory of Balanced Government," in *The Constitution Revisited,* ed. Conyers Read (New York, 1938), p. 44–45; Corinne C. Weston, *English Constitutional Theory and the House of Lords, 1556–1832* (London and New York, 1965), pp. 10–11; Lawrence Stone, *The Causes of the English Revolution, 1529–1642* (London, 1972), p. 104; and J. G. A. Pocock, *The Machiavellian Moment: Florentine Political Thought and the Atlantic Republican Tradition* (Princeton, N.J.: 1975), p. 433.

17. "The King's Answer to the Nineteen Propositions, 18 June 1642," in *The Stuart Constitution 1603–1688: Documents and Commentary,* ed. J. P. Kenyon, (Cambridge, 1966), p. 21.

18. John Milton, *Prose Works,* ed. J. A. St. John, 5 vols., (London, 1848–53), 2:125.

19. Ibid., 2:408.

20. Wilbur Cortez Abbot, ed., *The Writings and Speeches of Oliver Cromwell,* 4 vols. (Cambridge, 1937), 3:588 and 4:417, 418.

21. J. G. A. Pocock, ed., *The Political Works of James Harrington* (Cambridge, 1977), p. 405 (similarly, p. 164).

22. Ibid., p. 21.

23. Jacob Viner, "English Theories of Foreign Trade before Adam Smith," *Journal of Political Economy* 38 (1930): 249–301, 404–57, esp. p. 257; see also W. H. Price, "The Origin of the Phrase 'Balance of Trade,'" *Quarterly Journal of Economics* 20 (1905): 157ff.; and Edgar A. J. Johnson, *Predecessors of Adam Smith* (1937; New York, 1965).

24. Edward Misselden, *The Circle of Commerce* (London, 1623, repr. in Amsterdam and New York, 1969), pp. 116–17.

25. Francis Bacon, Letter to the Duke of Buckingham, August 1616, in *The Works,* ed. Spedding, Ellis, and Heath, 13:49, 22.

26. Sir James Steuart, *Principles of Political Economy* (1767), quoted in Johnson, *Predecessors of Adam Smith,* p. 308.

Chapter 8. Attraction and Repulsion

1. Woodrow Wilson, *Constitutional Government* (New York, 1908), p. 54, quoted in James A. Robinson, "Newtonianism and the Constitution," *Midwest Journal of Political Science* 1 (1957): 253.

2. For the background, see Derek T. Whiteside, "Before the *Principia*: The Maturing of Newton's Thoughts on Dynamical Astronomy, 1664–1684," *Journal for the History of Astronomy* 1 (1970): 5–19, esp. pp. 13–15; E. J. Aiton, *The Vortex Theory of Planetary Motions* (London and New York, 1972), esp. pp. 90–103; and the bibliography in I. B. Cohen, "Newton, Isaac," in *Dictionary of Scientific Biography* (New York, 1974), 10:98–99.

3. Isaac Newton, *The Correspondence,* ed. H. W. Turnbull (Cambridge, 1960), 2:307.

4. Ibid., 2:361.

5. Isaac Newton, *Mathematical Principles of Natural Philosophy,* ed. W. David, trans. Andrew Motte, 3 vols. (London, 1803), 1:2, 3.

6. For details, see chap. 3, pp. 97–98, above.

7. John Harris, *Lexicon Technicum,* 2d ed., 2 vols. (London, 1708), vol. 1 s.v. planets.

8. William Derham, *Astro-Theology* (London, 1715), p. 148.

9. J. T. Desaguliers, *Physico-Mechanical Lectures* (London, 1717), p. 10.

10. Comets and Planets in their Ellipses, move from the *Aphelion* (or greatest

distance from the Sun) to the *Perihelion,* (or nearest distance from the Sun) with an accelerated Motion; and from the Perihelion to the Aphelion, with a Motion uniformly diminish'd: The Attraction of the Sun, first accelerates the Motion, by conspiring with its Direction, then retards it, by drawing counter to it.

6. The Reason why a Planet, or Comet, does not fall into the Sun, when nearest to it, is, that the Centrifugal Force encreases in proportion to the square of the acquir'd Velocity; and the Reason that a Planet, or Comet does not go off, and leave the Sun, when at its Aphelion it is least attracted, is, that the Centrifugal Force diminishes in proportion to the Square of the diminish'd Velocity. (Ibid., p. 12)

11. For a sampling of representative titles, see Colin Maclaurin, *An Account of Sir Isaac Newton's Philosophical Discoveries,* ed. L. L. Laudan (New York and London, 1968), editor's introduction, p. xi.

12. John Rowning, *A Compendious System of Natural Philosophy,* 2 vols. (London, 1744), 1:iii.

13. Ibid., 1:iv–v.

14. On Gowin Knight, see Marie Boas, "The Establishment of the Mechanical Philosophy," *Osiris* 10 (1952): 523–24; and Robert E. Schofield, *Mechanism and Materialism: British Natural Philosophy in the Age of Reason* (Princeton, N.J., 1970), p. 175.

15. Boas, "The Establishment of the Mechanical Philosophy," pp. 509 and 515.

16. Walter Moyle, *Democracy Vindicated: An Essay on the Constitution and Government of the Roman State* (ca. 1699), ed. John Thelwell (Norwich, 1796), in Caroline Robbins, ed., *Two English Republican Tracts* (Cambridge, 1969), p. 63.

17. Henry Saint-John, First Viscount Bolingbroke, *The Works,* 4 vols. (London, 1844), 2:85.

18. Adam Smith, *The Wealth of Nations,* ed. Edwin Cannan (New York, 1937), p. 58.

19. J. T. Desaguliers, *The Newtonian System of the World: The Best Model of Government, An Allegorical Poem* (Westminster, 1728), p. v.

20. Ibid., pp. 30, 34.

21. Samuel Bowden, "A Poem Sacred to the Memory of Sir Isaac Newton," in *Poetical Essays on Several Occasions,* 2 vols. (London, 1733, 1735), 2:3.

22. Anon., *Occasional Reflections on the Importance of the War in America, and the Reasonableness of Supporting the King of Prussia* (London, 1758), p. 58, quoted in M. S. Anderson, "Eighteenth-Century Theories of the Balance of Power," in R. Hatton and M. S. Anderson, eds., *Studies in Diplomatic History* (New York, 1970), p. 189.

23. Edmund Burke, *Reflections on the Revolution in France* (London and New York, 1964), p. 33.

24. Ahlrich Meyer, "Mechanische und Organische Metaporik Politischer Philosophie," *Archiv für Begriffsgeschichte* 13 (1969): 175 ff.

1. See chap. 7, p. 142.

2. Anon., *Natural Reflections upon the Present Debates about Peace and War* (London, 1712), pp. 61–62, quoted in M. S. Anderson, "Eighteenth-Century Theories of the Balance of Power," *Studies in Diplomatic History,* ed. R. Hatton and M. S. Anderson (New York, 1970), p. 187.

3. Henry Saint John, First Viscount Bolingbroke, *The Works,* 4 vols. (London, 1844), 2:291.

4. Anon., *Occasional Reflections on the Importance of the War in America* (London, 1758), p. 58, quoted in Hatton and Anderson, eds., *Studies in Diplomatic History,* p. 189.

5. Jean Jacques Rousseau, *A Project for Perpetual Peace* (London, 1761), pp. 12–13, quoted in Hatton and Anderson, *Studies in Diplomatic History,* p. 189.

6. "A lasting general peace through the so-called balance of the powers in Europe is a mere phantom, like Swift's house which was built so perfectly according to all laws of equilibrium that it collapsed as soon as a sparrow landed on it" (Immanuel Kant, *Vom Verhältnis der Theorie zur Praxis im Völkerrecht,* in *Kleine Schriften* [1793], p. 112, quoted in A. Demandt, *Metaphern für Geschichte* [Munich, 1978], p. 307). Earlier J. H. G. von Justi had already published a book named *Die Chimäre des Gleichgewichts von Europa* (Altona, West Germany, 1758).

7. When Friedrich von Gentz, a Prussian political writer, observed in 1806 the continuously fluctuating character of the balance and its need to be stabilized by human hand, he was only repeating what Bolingbroke had said seventy years before (see n. 3 in this chapter):

> It perhaps could have been with more propriety called a system of *counterpoise.* For perhaps the highest of its results is not so much a perfect *equipoise* as a constant alternate oscillation in the scales of the balance, which, from the application of *counterweights,* is prevented from ever passing certain limits. (Friedrich von Gentz, *Fragments upon the Balance of Power in Europe* [London, 1806], p. 63n, quoted in Hatton and Anderson, *Studies in Diplomatic History,* p. 188)

8. Jonathan Swift, *A Discourse of the Contests and Dissentions in Athens and Rome* (London, 1701), p. 84.

9. [Barrington], *The Interest of England Considered in Respect to Protestants Dissenting from the Established Church* (1702), pp. 29–30, quoted in John A. W. Gunn, *Politics and the Public Interest in the Seventeenth Century* (London, 1969), p. 203.

10. Anon., *Two Essays on the Balance of Europe* (London, 1720), p. 774, quoted in Hatton and Anderson, *Studies in Diplomatic History,* p. 184.

11. J. Campbell[?], *Memoirs of the Duke of Ripperda* (London, 1740), appendix, p. 357, quoted in Hatton and Anderson, *Studies in Diplomatic History,* p. 185.

12. Daniel Defoe, *An Argument Showing that a Standing Army, with Consent*

of Parliament, Is not Inconsistent with a Free Government (1698), quoted in John G. A. Pocock, *The Machiavellian Moment* (Princeton, N.J., 1975), pp. 432–33.

13. Henry Saint John, First Viscount Bolingbroke, *Remarks on the History of England* (1730), in *The Works,* 1:306–7.

14. Sir William Blackstone, *Commentaries on the Laws of England,* ed. St. George Tucker, 5 vols. (Philadelphia, 1803, repr. South Hackensack, N.J., 1969), 1:51–52.

15. Walter Moyle, *Democracy Vindicated: An Essay on the Constitution and Government of the Roman State* (ca. 1699), in Caroline Robbins, ed., *Two English Republican Tracts* (Cambridge, 1969), pp. 244, 106.

16. Swift, *A Discourse of the Contests and Dissentions in Athens and Rome,* pp. 84–85.

17. See Corinne C. Weston, *English Constitutional Theory and the House of Lords, 1556–1832* (London, 1965), p. 87.

18. W. A. Craigie and J. R. Hulbert, *A Dictionary of American English on Historical Principles,* 4 vols. (Chicago, 1960), s.v. checks and balances (1:474).

19. Henry Saint John, First Viscount Bolingbroke, *A Dissertation upon Parties,* in *The Works,* 2:85.

20. [Gregory Sharpe], *A Short Dissertation upon that Species of Misgovernment Called an Oligarchy* (London, 1748), quoted in Stanley Pargellis, "The Theory of Balanced Government," in *The Constitution Reconsidered* (1938), ed. Conyers Read (New York, 1968), p. 44.

21. [Thomas Pownall], *A Treatise on Government being a Review of the Doctrine of an Original Contract* (London, 1750), p. 27.

22. Ibid., pp. 28–29.

23. David Hume, *Essays Moral, Political, and Literary,* ed. T. H. Green and T. H. Grose, 2 vols. (London, 1898), 1:100, 99.

24. Blackstone, *Commentaries,* 1:52.

25. Ibid., 2:154–55.

Chapter 10. Self-regulation in Economic Thought

1. See for example, Joseph Alois Schumpeter, *History of Economic Analysis* (New York, 1954), pp. 369–72.

2. Ibid., pp. 178, 196, 216, 243–49, 302, 353–54, 366, and 369; and Jacob Viner, *Studies in the Theory of International Trade* (1937; rep. New York, 1965), pp. 74–87.

3. Isaac Gervaise, *The System or Theory of the Trade of the World* (London, 1720), in Leonard Silk, ed., *The Evolution of Capitalism: Mercantilist Views of Trade and Monopoly* (New York, 1972).

4. Silk, p. 7.

5. Ibid., pp. 10 and 15.

6. Ibid., p. 5.

7. Ibid.

8. Richard Cantillon, *Essai sur la nature du commerce en général* (London,

1755), trans. and ed. Henry Higgs (New York, 1964), also excerpted in Ronald L. Meek, ed., *Precursors of Adam Smith, 1750–1775* (London, 1973).

9. Higgs, p. 27.

10. Ibid., p. 156.

11. Ibid., pp. 68–69.

12. David Hume, "Of the Balance of Trade," in *Essays Moral, Political, and Literary,* ed. T. H. Green and T. H. Grose, 2 vols. (London, 1898), 1:330–45.

13. Ibid., pp. 330 and 331.

14. Ibid., p. 333.

15. Ibid.

16. Ibid., pp. 334 and 340.

17. Ibid., p. 341.

18. Adam Smith, *The Wealth of Nations,* Modern Library Edition, ed. Edwin Cannan (New York, 1937).

19. Ibid., pp. 627 and 607; see also p. 418.

20. Ibid., p. 456.

21. Ibid., pp. 626, 625, and 595.

22. Ibid., pp. 56 and 62.

23. Ibid., pp. 99 and 634.

24. Ibid., p. 635.

25. Ibid., p. 404.

26. Ibid., p. 651.

27. Ibid., p. 14.

28. Ibid., pp. 14 and 421.

29. *Ibidem,* p. 423.

30. Ibid., chap. 7, "Of the Natural and Market Price of Commodities," pp. 55–63.

31. Schumpeter, *History of Economic Analysis,* p. 189.

32. Smith, *Wealth of Nations,* p. 57.

33. Ibid., p. 56.

34. When the quantity of any commodity which is brought to market falls short of the effectual demand, all those who are willing to pay the whole value of the rent, wages, and profit, which must be paid in order to bring it thither, cannot be supplied with the quantity which they want. Rather than want it altogether, some of them will be willing to give more. A competition will immediately begin among them, and the market price will rise more or less above the natural price, according as either the greatness of the deficiency, or the wealth and wanton luxury of the competitors, happen to animate more or less the eagerness of the competition. Among competitors of equal wealth and luxury the same deficiency will generally occasion a more or less eager competition, according as the acquisition of the commodity happens to be of more or less importance to them. Hence the exorbitant price of the necessaries of life during the blockade of a town or in a famine. (Ibid.)

35. When the quantity brought to market exceeds the effectual demand, it cannot be all sold to those who are willing to pay the whole value of the rent,

wages and profit, which must be paid in order to bring it thither. Some part must be sold to those who are willing to pay less, and the low price which they give for it must reduce the price of the whole. The market price will sink more or less below the natural price, according as the greatness of the excess increases more or less the competition of the sellers, or according as it happens to be more or less important to them to get immediately rid of the commodity. The same excess in the importation of perishable, will occasion a much greater competition than in that of durable commodities; in the importation of oranges, for example, than in that of old iron. (Ibid., p. 57)

36. When the quantity brought to market is just sufficient to supply the effectual demand and no more, the market price naturally comes to be either exactly, or as nearly as can be judged of, the same with the natural price. The whole quantity upon hand can be disposed of for this price, and cannot be disposed of for more. The competition of the different dealers obliges them all to accept of this price, but does not oblige them to accept of less. (Ibid.)

37. If at any time it [the supply] exceeds the effectual demand, some of the component parts of its market price must be paid below their natural rate. If it is rent, the interest of the landlords will immediately prompt them to withdraw a part of their land; and if it is wages or profit, the interest of the labourers in the one case, and of their employers in the other, will prompt them to withdraw a part of their labour or stock from this employment. The quantity brought to market will soon be no more than sufficient to supply the effectual demand. All the different parts of its price will rise to their natural rate, and the whole price to its natural price. (Ibid.)

38. If, on the contrary, the quantity brought to market should at any time fall short of the effectual demand, some of the component parts of its price must rise above their natural rate. If it is rent, the interest of all other landlords will naturally prompt them to prepare more land for the raising of this commodity; if it is wages or profit, the interest of all other labourers and dealers will soon prompt them to employ more labour and stock in preparing and bringing it to market. The quantity brought thither will soon be sufficient to supply the effectual demand. All the different parts of its price will soon sink to their natural rate, and the whole price to its natural price. (Ibid., pp. 57–58)

39. Ibid., p. 58.
40. Ibid., p. 59.
41. Ibid., p. 58.
42. Ibid., chap. 8, "Of the Wages of Labour," pp. 64–86.
43. Ibid., pp. 79 and 80.
44. The relationship between the working population and the difference between wage level and subsistence level is further elaborated on pp. 85–86.
45. Ibid., p. 80.
46. Ibid.

47. Ibid., chap. 10, "Of Wages and Profit in the Different Employments of Labour and Stock," pp. 99–118.

48. Ibid., p. 99.

Chapter 11. Self-regulation and the Liberal Conception of Order

1. Margaret C. Jacob, *The Newtonians and the English Revolution 1689–1720* (Ithaca, N.Y., 1976), passim; see index s.v. self-interest.

2. Richard Hooker, *Of the Laws of Ecclesiastical Polity,* ed. John Keble, 3 vols. (New York, 1970), 1:221.

3. George Savile, First Marquess of Halifax, *The Character of a Trimmer,* in *The Complete Works,* ed. Walter Raleigh (New York, 1970), p. 103.

4. Ibid., p. 48.

5. Sir Charles Wolseley, *Liberty of Conscience, the Magistrate's Interest,* (London, 1668), in John A. W. Gunn, *Politics and the Public Interest in the Seventeenth Century* (London, 1969), pp. 167–68.

6. William Penn, *A Persuasive to Moderation* (1686), in *Select Works,* 4th ed., 3 vols. (London, 1825, repr. New York, 1971), 2:529.

7. Penn, *Select Works,* 2:505.

8. David Hume, *Essays Moral, Political, and Literary,* ed. T. H. Green and T. H. Grose, 2 vols. (London, 1898), 1:99.

9. Penn, *A Persuasive to Moderation,* 2:529.

10. See Gunn, *Politics and the Public Interest,* p. 182.

11. For more material on *concordia discors,* see ibid., pp. 167–70, 181–83, 202, 283–84, and 325–26.

12. Bernard de Mandeville, *The Fable of the Bees, or Private Vices, Public Benefits* (London, 1714), in Adam Smith, *The Wealth of Nations,* ed. Edwin Cannan (New York, 1937), introduction, pp. lii–liii.

13. Alexander Pope, *Essay of Man,* in *The Poems of Alexander Pope,* ed. John Butt (London, 1963), p. 515.

14. For particulars see, for example, Sidney Pollard, *The Genesis of Modern Management* (London, 1965), chap. 5, and E. P. Thompson, "Time, Work-Discipline, and Industrial Capitalism," *Past and Present* 38 (1967): 56–97.

15. Montesquieu, *Considérations sur les causes de la grandeur des Romans et de leur décadence* (1734; Paris, 1876), pp. 102–3.

16. Edmund Burke, *Reflections on the French Revolution* (1790), ed. Sidney Lee (London, 1923), p. 37.

17. "Die Gesamtbewegung dieser Unordnung ist ihre Ordnung. In dem Verlauf dieser industriellen Anarchie, in dieser Kreisbewegung, gleicht die Konkurrenz sozusagen die eine Extravaganz durch die andere aus" (Karl Marx, "Lohnarbeit und Kapital," in Karl Marx and Friedrich Engels, *Werke,* ed. Institut für Marxismus-Leninismus beim Z.K. der S.E.D., [Berlin, 1973], 6:405).

Chapter 12. Self-regulating Mechanisms in Practical Technology

1. The following is largely based on material in Otto Mayr, *The Origins of Feedback Control* (Cambridge, Mass., 1970).

2. See Mayr, *Origins,* chap. 2.

3. Hero of Alexandria, *Spiritualium liber,* first Latin translation by Federigo Commandino (Urbino, 1575); see also Mayr, *Origins,* pp. 46–48.

4. An early example is the steam engine by Thomas Savery, patented in 1698 and described in his booklet, *The Miners' Friend* (1702, repr. London, 1858). The legend for figure 2 in that book lists "A cistern with a buy [sic] -cock coming from the force pipe"; the figure itself does not show the buoy-cock (i.e., float valve). The device is shown, however, most clearly in the account of Savery's engine given by Robert Stuart, *Historical and Descriptive Anecdotes of Steam Engines* (London, 1829), 1:118 and plates 3 and 4. I owe this reference to Dr. Svante Lindquist, Stockholm. Another early application of the float valve is on the Newcomen-type steam engine built in 1723 in Vienna by Emanuel Fischer von Erlach and described by Ludolf von Mackensen in "Ein wiedergefundenes Zwischenglied in der Geschichte der Regelungstechnik," *Regelungstechnik und Prozess-Datenverarbeitung* 19 (1971): 425–28. For further material, see Mayr, *Origins,* chap. 7.

5. Mayr, *Origins,* chap. 6.

6. Ibid., chap. 8.

7. Hans Peter Münzenmayer, "Leibniz' Inventum Memorabile: Die Konzeption einer Drehzahlregelung vom März 1686," *Studia Leibnitiana* 8 (1976): 113–19.

8. Mayr, *Origins,* chap. 9.

9. Ibid., chap. 10.

10. Norbert Wiener, *Cybernetics, or Control and Communication in the Animal and the Machine* (Cambridge, Mass., 1948), pp. 11–12.

Illustration Credits

1–1 Reproduced, with permission, from a photo by Klaus Maurice, Munich.

1–2(A) Reproduced, with permission, from a photo by Klaus Maurice, Munich.

(B) Reproduced from R. P. Howgrave-Graham, "The Crowing Cock, Strasbourg," *The Watch and Clock Maker,* September 15, 1930, figure 85.

1–3 Reproduced from MS no. 172, figure 7, Eton College Library, with the permission of the Provost & Fellows of Eton College.

1–4 Reproduced from Denis Diderot and Jean Le Rond d'Alembert, *Encyclopédie,* Recueil de planches (Paris, 1765), vol. 8, plate 10.

1–5 Reproduced, with permission, from a photo by Bayerisches Nationalmuseum, Munich.

1–6 Reproduced, with permission, from a photo by Peter Frieß, Munich.

1–7 Reproduced from a woodcut by Tobias Stimmer, "Eigentliche Fuerbildung und Beschreibung dehs newen Kuenstrechen Astronomischen Urwercks," Strasbourg, 1574.

1–8(A) Reproduced, with permission, from a photo by the Smithsonian Institution, Washington, D.C.

(B) Reproduced, with permission, from a photo by Peter Harding, Tetbury, England.

1–9 Reproduced, with permission, from a photo by Kunsthistorisches Museum, Vienna.

1–10 Reproduced, with permission, from a photo by Photographie Giraudon, Paris.

1–11 Reproduced, with permission, from a photo by The National Gallery, London.

1–12 Reproduced, with permission, from a photo by The Chronological Collection of the Danish Kings at Rosenborg Castle, Copenhagen.

1–13 Reproduced, with permission, from a photo by Peter Frieß, Munich.

1–14 Reproduced, with permission, Bildarchiv Foto Marburg im Forschungsinstitut für Kunstgeschichte, Marburg.

1–15 Reproduced, with permission, from a photo by Astronomisch-Physikalisches Kabinett, Staatliche Kunstsammlungen, Kassel.

1–16 Reproduced, with permission, from a photo by Bayerisches Nationalmuseum, Munich.

2–1 Reproduced, with permission, from a photo by Bibliothèque Royale Albert Ier, Brussels.

2–2 Reproduced from MS Laud misc. 570, fol. 28, with permission of the Curators of the Bodleian Library, Oxford.

2–3 Reproduced from an engraving, "Temperantia," by Pieter Bruegel the Elder, circa 1560.

3–1 Reproduced from Christian Vurstisius, *Questiones novae in theoricas novas planetarum doctissimi mathematici Georgii Purbachii,* Basel, 1568, p. 210.

12–1 Reproduced from Hero of Alexandria, *Opera . . . omnia,* ed. and trans. W. Schmidt (Leipzig, 1899), vol. 1, fig. 75.

12–2 Reproduced from Hero of Alexandria, Spiritualium liber, trans. Federigo Commandino (Urbino, 1575), p. 69.

12–3 Reproduced from Robert Stuart, *Historical and Descriptive Anecdotes of Steam Engineers and Their Inventors and Improvers* (London, 1829), plate IV.

12–4 Reproduced from *Annuals of Science* 6 (1948) 40, figure 1.

12–5 Reproduced from *Annuals of Science* 6 (1948) 41, figure 2.

12–6 Reproduced from *Acta eruditorum Lipsiae* (1682), p. 106.

12–7 Reproduced from *Studia Leibnitiana* 8.1 (1976) 114, with permission of Franz Steiner Verlag, Wiesbaden.

12–8 Reproduced from British Patent (Old Series) no. 1628 (1787), of Thomas Mead, Regulator for Wind and Other Mills.

12–9 Reproduced from William Fairbairn, *Mills and Millwork* (London, 1861), section 1, figure 174.

12–10 Reproduced from Conrad Matschoss, *Die Entwicklung der Dampfmaschine* (Berlin, 1908), vol. 1, fig. 107.

Index

Alanus ab Insulis (Alain de Lille), 39
Animal automatism, 64–67, 69–70, 78–79, 89–92, 104, 107–8
Antikythera device, 5, 58
Aquinas, Thomas, 207 n. 33
Archimedes, 39, 58, 63, 111
Argument from design, 38–40, 47–49, 57, 68, 77, 78, 107, 116, 122–23; determinism vs. free will, 92–101; voluntarist-intellectualist debate, 48–49, 73, 93, 98, 123
Aristotelian scholasticism, 83–84
Aristotle, 102, 141
Arnobius Afer, 39
Astronomical clocks, 7–8, 10–13, 21, 58–59, figs. 1-3, 1-5, 1-6, 1-7, 1-12, 1-15
Atheism, 57, 92, 95
Attraction (gravitational) and repulsion, 151–54
Augsburg, as center of clockmaking, 9
Authoritarian conception of order, 119–21, 126, 139, 196
Automata: history of, 4, 7, 21–26, figs. 1-2, 1-12, 1-13, 1-14, 1-16; Descartes's interest in, 63–64; legends, 24–25, 63, 107; metaphors, chaps. 3–6, passim. *See also* Animal automatism; Program control

Bacon, Francis, 55, 56, 59, 142, 146
Balance metaphors, chaps. 7–11, passim; significance and early history, 140–41
Balance of power, 141–43, 156–59
Balance of trade, 146–47, 164–69, 172–73
Balanced government, 158–63
Beaumont, Francis, 50, 212 n. 77
Berkeley, George, 96, 126
Berthold of Freiberg, 32–33
Blackstone, Sir William, 159, 162–63
"Body politic" analogy, 102, 104–5
Bolingbroke, Lord. *See* Saint-John, Henry, First Viscount Bolingbroke
Bowden, Samuel, 153
Boyle, Robert, 54–55, 82, 83, 85–92, 94–96
Brahe, Tycho, 59
Brandes, Ernst, 109, 135, 240 n. 31, 248 n. 55
Brenner, Steffen, 12, fig. 1-6
Bright, Timothy, 42
Bruegel, Pieter, the Elder, 37, fig. 2-3
Burke, Edmund, 153, 189
Burnet, Thomas, 90, 95

Calcidius, 39
Camden, William, 142
Cantillon, Richard, 168–69
Caroline, Princess of Wales, 98
Cartesian philosophy. *See* Descartes
Cartwright, William, 50
Cassiodorus, 39
Centralistic command structures, 65–66, 69, 104, 112, 117–20
Champaigne, Philippe de, 19
Charles I, King of England, 144
Charles V, Emperor, 49
Checks and balances, 140–41, 158, 160–63
Cheyne, George, 96, 126
Christine de Pisan, 35, 36, fig. 2-2, 40

Cicero, 39, 58
Clarke, Samuel, 73, 98–101, 122, 125–26
Clement of Rome, Saint, 39
Clock, mechanical, chaps. 1–6, passim; history: First phase, weight-driven, 6–8; Second phase, spring-driven 8–12; Third phase, pendulum and balance spring, 12–14; origins and invention, 3–5; hypothesis of eastern origins, 5; as symbol of aristocratic rank, 17, figs. 1-9, 1-10, 1-11; uses, chap. 1, passim; summary, 26
Clockmaker-God. *See* Argument from design
Clockmakers, 6, 8–9, 12–14
Clock metaphor, chaps. 2–6, passim; evaluation and summary, 115–19; methodology, 28–30. *See also* Metaphors
Clocks and automata, historical significance, xvii
Comenius, John Amos, 42
Competition, 127, 175
Concordia discors, 185–86
Condillac, Etienne Bonnot de, 109
Constitutional government (or monarchy), 127, 129, 158, 159, 162, 163, 183. *See also* Balanced government; Limited government; Mixed government
Contarini, Gasparo, 143
"Contingent," Christian Wolff's use of the term, 76
Copernican system, 59
Cotes, Roger, 84, 126
Cromwell, Oliver, 125, 145
Crusius, Christian August, 131
Cudworth, Ralph, 95
Cybernetics, 65, 195

Dafydd ap Gwilym, 15, 50
Dante Alighieri, 30–32, 40
Darwin, Charles, 123
Dasypodius, Conrad, 10, 12, fig. 1-7
Davenant, William, 43
Davies of Hereford, John, 52
Defoe, Daniel, 158–59
Dekker, Thomas, 212 n. 79
Derham, William, 96, 150
Desaguliers, J. T., 151, 153
Descartes, René, 56, 62–71, 77, 81–82, 87, 90–92, 104
Design argument. *See* Argument from design
Determinism, 71–78, 92–101, 104, 107, 118–19, 125, 130–33. *See also* Free will; Argument from design
Diderot, Denis, 78–80, 110, 113, 130
Digby, Sir Kenelm, 82, 91, 93
Dionysius Areopagita, 39
Dondi, Giovanni de', 7–8, fig. 1-3, 59
Donne, John, 42, 44, 46, 50
Drebbel, Cornelis, 190–91, 193, figs. 12-4, 12-5
Dynamic equilibrium, 139, chap. 8, passim, 161, 187. *See also* Self-regulation

Economic liberalism, chap. 10, passim, 174
Engine, ambivalence of word, 124
English ambivalence about clocks, 49–53, 123–26
English clocks, 14, 128
Equilibrium, chaps. 7–11, passim. *See also* Balance metaphors
Escapement, verge-and-foliot, 5, 31, 34
Euler, Leonhard, 107, 131–32
Eusebius, church father, 58

Fan-tail, 193, 197, fig. 12-9
Father Time, 41
Feedback mechanisms, xv–xvi, chap. 12
Feedback principle, xvi, xviii, 120, 121, 139, 166–68, 176–80, 187. *See also* Self-regulation; Dynamic equilibrium
Fénélon, François de Salingae de la Mothe, 69
Fichte, Johann Gottlieb, 134, 136
Figure clocks, 21–24. *See also* Automata
Firmicus Maternus, 39
Fletcher, John, 50, 51, 212 n. 77
Float-level regulator, 190, figs. 12-1, 12-2, 12-3
Fontenelle, Bernard de, 70
Frederick II, "the Great," King of Prussia, 107–9, 134
Freedom, 76, 77, chap. 6, passim, 109, 139, 172–73, chap. 11, passim. *See also* Economic liberalism; Free will; Liberal conception of order; Liberty
Free will, 65–66, 72, 92–101, 130. *See*

also Determinism; Argument from design
Friedrich Wilhelm I, "the Soldier King," King of Prussia, 76
Froissart, Jean, 33–35, 40
Fusee, 8–9, fig. 1-4

Galilei, Galileo, 14, 56, 60, 65, 214 n. 12
Gear trains, early technology of, 5
George I, King of England, 98
German clocks, negative image in England, 49–52
Gervaise, Isaac, 165–68
Geyger, Johannes, 42, 45, 48
Gilbert, William, 59
Giovanni da Fontana, 40, 207 n. 33
Glanvill, Joseph, 84, 90, 95
Goethe, Johann Wolfgang, 109, 240 n. 30
Gottsched, Johann Christoph, 57, 130
Governor (speed regulator), 193–95, 198, fig. 10-12
Granada, Luis de, 47, 45
Grosseteste, Robert, 39
Guevara, Antonio de, 43
Guicciardini, Francesco, 142
Gustavus Adolphus, King of Sweden, 46

Habrecht, Isaak and Josias, 12, 22, figs. 1-7, 1-12
Hale, Mathew, 83, 90
Halifax, Lord. *See* Savile, George, First Marquess of Halifax
Hardenberg, Friedrich von ("Novalis"), 136
Harrington, James, 145–46
Harris, John, 150
Harrison, John, 128
Harsdörffer, Georg Philipp, 44
Hegel, Georg Wilhelm Friedrich, 134, 136
Helvétius, Claude Adrien, 113
Henry of Langenstein (or, of Hesse), 40
Herbert, George, 46
Hero of Alexandria, xvi, 58, 190–91, figs. 12-1, 12-2
Hipparchus, 58
Hobbes, Thomas, 56, 104–5, 127, 186, 214 n. 12
d'Holbach, Paul Henri Thiry, 113, 130, 243 n. 59
Hooke, Robert, 14, 85, 89
Hooker, Richard, 182
Horological Revolution, 14–17, fig. 1-8, 128
Horologium devotionis, Horologium sapientiae, 32–33, fig. 2-1
Hour, 3–4, 15–16
Hume, David, 122, 136, 162, 169–71, 185
Huggens, Christian, 14, 68

Invisible Hand, 165, 175
Intellectualism. *See* Argument from design

Joannes Cassianus (John Cassian), church father, 32
"Jacks" (automaton figures), 50
Jonson, Ben, 50, 51
Justi, Johann Heinrich Gottlieb von, 111

Kameralism, 111, 112
Kant, Immanuel, 57, 132–34, 136, 157
Keill, John, 56, 151
Kepler, Johannes, 60–61
King, Henry, Bishop, 46
Knight, Gowin, 152

Lactantius, 39
Laissez faire, 165. *See also* Economic liberalism
La Mettrie, Julien Offray de, 79, 107, 130
Langebek, Detlev, 102
Le Grand, Anthony, 91
Lehmann, Christoph, 43, 44
Leibniz, Gottfried Wilhelm, 55, 70–73, 77, 98–101, 122, 131, 133, 191
Leibniz-Clark debate, 72–73, 77, 98–101, 122
Liberal conception of order, 128, 139, 155, 181–89, 196–99. *See also* Authoritarian conception of order; Self-regulation
Liberalism, liberty, 78, 122–36, chaps. 7–12, passim. *See also* Economic liberalism; Freedom; Liberal conception of order
Limited government (or monarchy), 127, 144, 146, 153, 158. *See also* Balanced government; Constitutional government; Mixed government

Lipsius, Justus, 44, 102
Locke, John, 84–85, 86–87, 89, 105
Lohenstein, Daniel Casper von, 106
Lucretius, 39
Luis de Granada, 45

Machina mundi, 39
Machine, ambivalence of word, 124
Machine analogy, chaps. 3–6, passim, esp. 56–58, 135–36
Maclaurin, Colin, 126
Magic, 24, 26, 55
Malebranche, Nicolas, 68–69
Mandeville, Bernard de, 165, 185–86
Marx, Karl, 135, 189, 248 n. 57
Materialism, 57, 79, 92
Mechanical, ambivalence of word, 55, 124–25
Mechanical clock. *See* Clock, mechanical
Mechanical philosophy, chaps. 3–6, passim, esp. 54–62
Mechanical world models, 58–61
Mercantilism, 106, 147, 164, 171–72
Mesure (*misura, mâze*), 34–35. *See also* *Temperantia*
Metaphors: abundance in sixteenth- and seventeenth-century literature, 41; discouraged by Scientific Revolution, 61; indicating group mentalities, 28–30; of order and authority, 115–19. *See also* Animal automatism; Balance metaphors, "Body politic" analogy; Clock metaphor; Machine analogy; Organic imagery; Ship-of-state imagery; State, metaphors for the
Middleton, Thomas, 50–51
Mill, John Stuart, 123
Milton, John, 128, 144–45
Miracles in deterministic philosophy, 75
Misselden, Edward, 146
Mixed government (or constitution), 141, 143–46, 161. *See also* Balanced government; Constitutional government; Limited government
Monads, 72–73
Montesquieu, Charles Louis de, 129, 189
More, Henry, 90–91, 93
Mornay, Philippe de, 47
Moser, Friedrich Karl von, 111
Moyle, Walter, 105, 152, 159
"Muddling through," the ethos of, 120, 188
Müller, Adam Heinrich, 134, 248 n. 56
Murillo, Bartolomé Esteban, 20, fig. 1-11
Musical capabilities in clocks and automata, carillons, 7, 10, 23, fig. 1-13, 109, 132, 136, 247 n. 43, 248 n. 59

Nashe, Thomas, 50
Nature imagery. *See* Organic imagery
Newton, Sir Isaac, 73, 77, 97–101, 122, 125, 148–50
Newton's "clockwork universe," 98
Newtonian System, Newtonianism, 148–54
Nicole Oresme, 38–39, 40
Nikolaus von Cues, 40, 207 n. 33
Norden, John, 48
Novalis. *See* Hardenberg, Friedrich von

Order, conceptions of, 119–21, 139, 181. *See also* Authoritarian conception of order; Liberal conception of order
Organic imagery, 136, 140. *See also* Animal automatism; "Body politic" analogy
Overbury, Sir Thomas, 142

Paley, William, 123
Papin, Denis, 191
Pascal, Blaise, 67–68
Patrick, Simon, 83
Paxton, James, 123
Pelham, Sir William, 46
Pendulum clock, 14
Penn, William, 105, 184–85
Pfeiffer, Johann Friedrich von, 111–12
Physiocrats, 165
Picinelli, Filippo, 41
Pineal gland (soul), 65, 78
Planetariums in antiquity, 58
Poisson, N. I., Father, 63
Polybius, 141
Pope, Alexander, 186
Posidonius of Apamea, 58
Power, Henry, 83, 90, 93
Pownall, Thomas, 161–62
Preestablished harmony, 71–72
Pressure regulator (safety valve), 191, 194, fig. 12-6
Profit motive, 164. *See also* Self-interest
Program control, 4, 7, figs. 1-2, 1-13, 1-14, 21, 23, 66, 109, 118

Ptolemy, Ptolemaic system, 58, 59
Punctuality, 17

Quadrivium, 10

Rabelais, François, 15, 41, 203 n. 21
Rheticus, Joachim, 47
Richard of Wallingford, 7–8, 59
Ripa, Cesare, 43
Robinson, John, 48–49
Rohan, Henri Duc de, 102
Rousseau, Jean Jacques, 109–10, 112, 130, 157
Rowning, John, 151–52

Saavedra Fajardo, Diego de, 103
Sacrobosco, John, 39
Saint-John, Henry, First Viscount Bolingbroke, 152–53, 156–57, 159–61
Santa Cruz, Antonio Ponce de, 103
Sapientia, in *Horologium sapientiae,* 32, 33, fig. 2-1, 35
Savile, George, First Marquess of Halifax, 105, 127, 183–84, 239 n. 13
Schiller, Friedrich, 136
Schlözer, August Ludwig von, 112
Schmidt, Ludwig Benjamin Martin, 111
Schoonhovius, Florentius, 103
Schröder, Wilhelm von, 106
Self-interest, 162, 164, 174–75, 182, 185–86
Self-regulation, 139, 155, 157, 164–80, 186–88. *See also* Dynamic equilibrium; Feedback principle; Liberal conception of order
Shakespeare, William, 50–52, 124, 212 n. 83
Sharpe, Gregory, 161
Ship-of-state metaphor, 102
Sidney, Algenon, 127–28
Smith, Adam, 14, 106, 135, 153, 164–65, 171–80
Soul, 65, 69, 77–79, 91–92
Speed regulator, 191–98, figs. 12-7, 12-8, 12-10
Spinoza, Baruch, 68
Sprat, Thomas, 61, 82, 84, 127
Spring-driven clocks, 8–12
State, metaphors for the, 102. *See also* Metaphors
Steuart, Sir James, 147
Strasbourg cathedral clock, 7–8, 10, figs. 1-2, 1-7, 1-12, 59, 63, 82, 86, 87, 94
Stubbe, Henry, 95
Suckling, Sir John, 212 n. 86
Sundial, 4
Supply and demand, law of, 175–78
Suso, Henricus (Heinrich Seuse), 32–33, 43
Swift, Jonathan, 158, 160
Sydenham, Thomas, 56, 86
System, 79, 117, 161

Technology, general character of, xv, xvii
Temperantia (temperance), 34–38, figs. 2-2, 2-3
Theological aspects of mechanistic philosophy, 57, 92
Thermostat, 190–91
Thomson, James, 160
Time, timekeeping, 3–27, 41, 51, 53, 128, 212 n. 83, 248 n. 59. *See also* Father Time; Hour; Punctuality
Trimmer, 183–84
Tymme, Thomas, 45–46

Ussher, James, Bishop, 93
Utrecht, Treaty of, 142

Vauçanson, Jacques de, 133
Verge-and-foliot escapement, 5
Vicious circle, 187
Visconti, Gaspare, 207 n. 33
Vitruvius, 58
Voltaire, 77, 107, 129, 130
Voluntarism. *See* Argument from design

Waller, Edmund, 142
Water clock, 4
Webster, John, 43, 46, 57
Whiston, William, 150
Wiener, Norbert, 65, 195
Wilkins, John, 47
William III, King of England, 142
Willis, Thomas, 84
Wilson, Woodrow, 148
Wolff, Christian, 73–77, 107, 130, 131
Wolseley, Sir Charles, 184
World Machine. *See* Argument from design; *Machina mundi*; Mechanical world models

JOHNS HOPKINS STUDIES IN THE HISTORY OF TECHNOLOGY

General Editor: Thomas P. Hughes

Books in the Series

The Mechanical Engineer in America, 1839–1910: Professional Cultures in Conflict, by Monte Calvert

American Locomotives: An Engineering History, 1830–1880, by John H. White, Jr.

Elmer Sperry: Inventor and Engineer, by Thomas Parke Hughes (Dexter Prize, 1972)

Philadelphia's Philosopher Mechanics: A History of the Franklin Institute, 1824–1865, by Bruce Sinclair (Dexter Prize, 1975)

Images and Enterprise: Technology and the American Photographic Industry, 1839–1925, by Reese V. Jenkins (Dexter Prize, 1978)

The Various and Ingenious Machines of Agostino Ramelli, edited by Eugene S. Ferguson, translated by Martha Teach Gnudi

The American Railroad Passenger Car, New Series, no. 1, by John White, Jr.

Neptune's Gift: A History of Common Salt, New Series, no. 2, by Robert P. Multhauf

Electricity before Nationalisation: A Study of the Development of the Electricity Supply Industry in Britain to 1948, New Series, no. 3, by Leslie Hannah

Alexander Holley and the Makers of Steel, New Series, no. 4, by Jeanne McHugh

The Origins of the Turbojet Revolution, New Series, no. 5, by Edward W. Constant II (Dexter Prize, 1982)

Engineers, Managers, and Politicians: The First Fifteen Years of Nationalised Electricity Supply in Britain, New Series, no. 6, by Leslie Hannah

Stronger Than a Hundred Men: A History of the Vertical Water Wheel, New Series, no. 7, by Terry S. Reynolds

Authority, Liberty, and Automatic Machinery in Early Modern Europe, New Series, no. 8, by Otto Mayr

The Johns Hopkins University Press

AUTHORITY, LIBERTY, AND AUTOMATIC MACHINERY IN EARLY MODERN EUROPE

This book was set in Simoncini Garamond text and display type by the Composing Room of Michigan, Inc., from a design by Chris L. Smith. It was printed on 50-lb. Booktext Natural paper and bound in Arrestox A by BookCrafters.

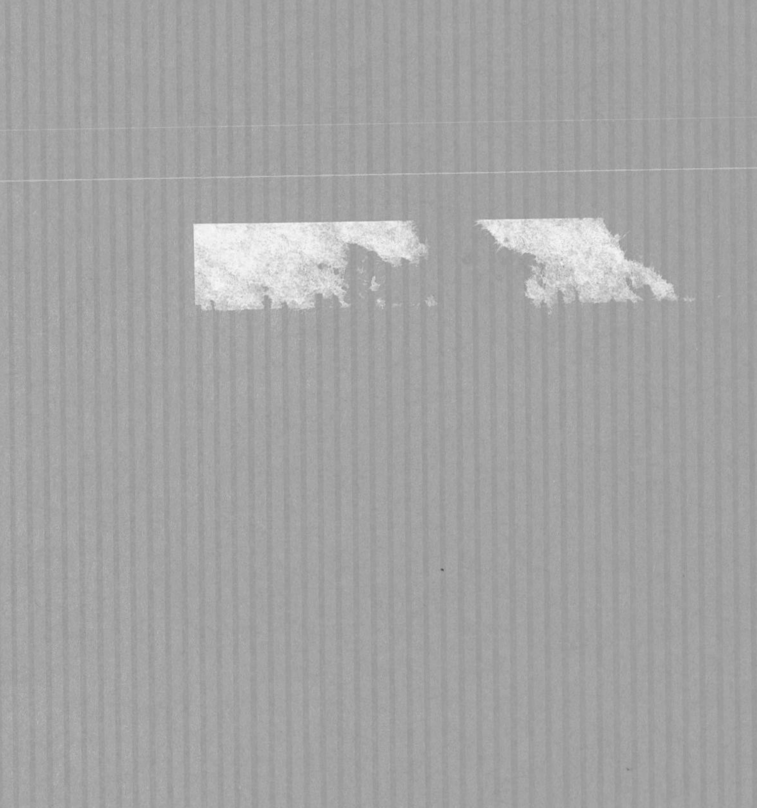